Climbing the Corporate Ladder–Safely!

Mark D. Hansen

American Society of Safety Engineers ❖❖ Des Plaines, Illinois USA

Climbing the Corporate Ladder–Safely!

Library of Congress Cataloging-in-Publication Data

Hansen, Mark D.

Climbing the corporate ladder–safely! / Mark D. Hansen.
p. cm
ISBN 978-1-885581-57-0 (alk. paper)
1. Career development. 2. Industrial safety–Vocational guidance. 3. Health–Vocational guidance. 4. Environmentalists–Vocational guidance. I. Title.

HF5381.H1362 2011
650.14–dc22 2011002137

Managing Editor: Michael Burditt, ASSE
Associate Editor: Jeri Ann Stucka, ASSE
Copy editing, page design, and layout: Nancy Kaminski
Cover: Troy Courson, Image Graphics

Printed in the United States of America
17 16 15 14 13 12 11 7 6 5 4 3 2 1

Dedication

Throughout my career, I have been greatly blessed. I have had good and bad experiences, and both have molded me into what I am today. I have worked diligently to protect what is important to me and maintain the priorities of life–God, family, and work.

I thank God for blessing me with the talent to write and the desire to put my thoughts into words so others can learn, laugh, and benefit from them.

I thank my dedicated and lovely wife of 23 years, Kathryn, and my beautiful daughter, Hannah, who have been the apples of my eye and my fortitude throughout my life.

I thank my employer (and extended family), Range Resources Corporation. It is a joy to work with such great people.

Table of Contents

Foreword

After writing *Out of the Box* and its recent update, I had another epiphany: once you get to your next career goal, or close to it, how do you stay there? How do you move up without losing the ground you've gained? You've developed a plan and painstakingly followed it, step by step. You've flexed with the plan as situations fluxed. Sometimes you went up, sometimes you went sideways, and sometimes you had to take a step back in order to take three steps forward. You've navigated the pitfalls and leveraged your successes. You are where you want to be—for now.

You are successful and well-placed on the corporate ladder, perhaps you are even the senior SH&E person in your organization. All of that was clearly difficult to achieve. However, what lies ahead of you is perhaps your most daunting challenge, staying on the ladder and inching your way up safely without sliding down or falling off.

At this level, you cannot hide. You are a proverbial big fish in a little pond. Even worse, the fish pond is more like a fish bowl, with everyone watching your every move, sometimes under a microscope. Your mistakes are magnified and your successes muted. You are the company subject matter expert. To the company, you know everything about SH&E, whether you really do or not. If you don't, you need to know how to find people who do, fast. You can either climb up the ladder with ease or struggle to sustain your position. Within these pages you will see some of my insights to career success in this area. At the end of every chapter, you will see a quick list of success tips to help you summarize the chapter.

When the time comes, you may need to hire SH&E professionals—good ones, not just warm bodies. If you hire someone who turns out to be a bust, you get the blame. If your hire succeeds, he or she gets the credit. It isn't fair, but then, no one said it would be. These are the facts. So hiring SH&E professionals can easily make n or break your reputation. Approach it with care. Handle it correctly and you can establish yourself as a talent finder. If you do well with hiring, you can move up the ladder with relative ease. Fail at it and everyone will question your judgment. Your candidates will undergo scrutiny such that it appears to be unfair, all due to past performance on your part. If you fail here, moving up the ladder may very well be questioned.

If you are sought after, you will be in the driver's seat when climbing the ladder. What this means is that sister companies will see your success and want the same results. I've actually seen really good performers cross industries, all based on demonstrated performance. They may come courting you. This is a great place to be, but be careful. If you have a great job you love, you are in the minority. If you are in a job you abhor, no amount of money is enough to make it tolerable. Evaluate opportunities carefully. You could easily move up and out (of the company) or just be happy where you are. One of the keys to success is knowing you are where you want to be, and being happy there. Just enjoy it. Many fail to do just that.

You may be looking for a job to move you up the ladder. In that case, you need to differentiate your resume from the rest of the pack. Many resumes look the same. How do you make yours different to get prospective employers to call you and *not* the other person? I've presented some key tips here for you to ponder. Writing a resume should be something we do often, not just when you need a job. Keep it up to date to capture recent accomplishments. If you don't, you risk forgetting that one accomplishment that may catch the eye of the resume reader. I've also included interviewing issues for SH&E professionals. I've interviewed quite a bit over the years. I've done some interviews well, while others were miserable failures. Hopefully, you can learn from my mistakes and gain from my lessons learned.

On the other hand, if you are like some, you are experiencing the downside of employment and looking for a job, or if employed, a better job. What I've included here are some principles, experiences, and strategies when looking for that job. Do this right and you will be well on your way up the ladder. Do it poorly and you could be destined for a less-than-desirable job further *down* the ladder.

What if you wake up one day and you are not on a ladder anymore—you're out of a job? Or the ladder is suddenly much steeper—you are expected to work twice as hard for the same pay, and you want out? In the chapter, "Corporate Survival", I discuss how to know when it is time to leave the company, depending on the economy, your position, and stature. Also discussed is surviving during a downsizing. I've been fortunate to go through all aspects of mergers and acquisitions. As a result, I hope you can turn my experiences into some gems for surviving these changes. Often they can come as a surprise, and other times they are well anticipated. Being informed and in the know will help you prepare in moving up, down, or sideways, depending on which end of the transaction you are on.

Politics is one place many SH&E professionals either fail or perform in a lackluster manner, stifling their ability to move up the ladder. Many view this as a game. If you view politics in this way, my word of caution to you is that it is a dangerous game. As an engineer, my approach was to develop a political roadmap for dealing with company politics. I got the idea from one of my early bosses while working in the aerospace industry. I found this industry steeped with politics. This was due to the fact that they hired many retired military people who were experts at the game and experts at excluding non-military people and protecting themselves. I had to devise some kind of strategy to understand who was who. The organizational chart was only the tip of the iceberg. My first organizational chart helped me to determine who the real players were and the characteristics of the clique. It helped me understand and deal with the way the organization really functioned. I have also included safe politics as well as the politics of safety. It is difficult to acknowledge, but SH&E often becomes a political football with us in the middle. Dealing with politics well can move you up the ladder with more ease than you might otherwise be able to accomplish.

There is a lot of information in this book on dealing with bosses, because we can learn from both the good ones and the bad ones. Some are just good strategies to make life easier for you and your boss. Other strategies include how to cope with a bad boss. We've all had bad bosses who make us appreciate the good ones. It also gives us a great basis for being a good boss and not wanting to be a bad boss. In the end, we need to learn from both good and bad bosses. Knowing this and acting on it can move you up the ladder.

Similar to understanding politics, and equally as important, is understanding business. It almost doesn't seem fair. We view SH&E altruistically, but in reality companies must make money. Understanding business, and developing business strategies, can allow you to jump many rungs of the ladder to the very top very quickly. Not understanding business can relegate you to being an SH&E geek. I know it doesn't sound fair, but it is the cold hard truth of business. In this book you will find many strategies I have used to help me successfully scale the ladder to my current position. Let me remind you that these are only my experiences. I haven't seen or done everything, but I've seen a lot. I hope this will help you to scale the ladder and bypass my mistakes.

When talking about SH&E professionals, getting paid for what we do is a key component. Without adequate pay we would all be in poverty. Yes, we work in an altruistic discipline, but we deserve to make a decent living. Many SH&E professionals fail to realize this fact. It is evidenced in our inability to garner top dollar for our services. If you use the business approach to SH&E, you are somewhat better prepared to talk about salary to your employer. If not, you will get what they offer and won't negotiate. Much of moving up the ladder is about negotiating a top salary. If it weren't, you probably wouldn't be reading this book. Within these pages are some tips and strategies I've used successfully to get top dollar. I've done this while employed by the same employer, getting promoted, and while changing employers. A key issue here is being able to say no and walk away. Some of my best offers came a few months after I said no to a potential employer. Behind this is the patience to wait for the right time to negotiate. Clearly, if you want to move up the ladder, you must master the art of negotiation and be able to talk about your salary like a businessperson who understands SH&E, not like an SH&E geek.

One of the key components to moving up the ladder is knowing how to manage your time. Without time management, goal setting, and being able to delegate, we will be relegated to the category of also-rans. Good time management is key to good job performance and exceeding expectations by being consistently proactive. You may be labeled as a good professional, but you just don't have that extra something to take the company to the next level. All three—time management, goal setting, and being able to delegate—are intertwined. Done well, they will get you considered for the next level. Do it outstandingly and you may be a candidate for a quantum leap (something we would all want). There are a lot of books and magazine articles on these topics. However, SH&E is decidedly different. We are a 24/7 job. How do you manage your time when you are supposed to be available all the time? I have provided some skills to do just that. (What is truly amazing is that they don't teach this stuff anywhere.)

When it comes to email, I've literally had *e-nough*. Whatever happened to face-to-face conversations? Our job is building relationships, and it takes these types of conversations to be successful. Short of that, we rely heavily on the phone, specifically, the cell phone. We've been equipped with cell phones for over twenty years, but people still lean on email, instant messaging, text messages, and Twitter. It is unlikely that email will move you up the ladder, however—and this is a BIG however—it could easily move you down several rungs and potentially out the door. All of this with *one* lousy, short, email! In this book, I've presented some of the key pitfalls with email, such as what to say and what not to say, and how to say it. In the end, if it is a delicate matter, it is likely not appropriate for email.

Demonstrated leadership may not move you up the ladder as quickly as some of the others, but without it, you are destined for the trenches. Leadership takes boldness in a world of spin and blame. We must deliver the good news and the bad news. We have to manage our employees and our boss. Managing your employees is straightforward. Managing your boss is quite different. I know this sounds like manipulation, but it isn't. It's about your boss's expectations and personality style. Understand these and apply them, and you can get along with any boss, even a micromanager. (By the way, the worst ones are those who constantly tell you they are macromanagers, and then proceed to micromanage you.) Being able to manage well may help you move up the ladder. I've seen this through the years. This is why safety became the catch-all for many ancillary disciplines. The safety department evolved into the HSSEQ (Health,

Safety, Security, Environmental, and Quality) department. What's next? This is good for SH&E. How better to move up the ladder as you grow your responsibilities than by proving that can you can manage it well as your department grows.

In the end, I hope you are able to take this book, read it, digest it, and dog-ear it as a daily career guide. I hope you can use it for those personal issues you face in striving to achieve fulfillment in your career. I hope it can help you not only scale the ladder, but scale it safely throughout your career.

Best wishes and good reading.

–Mark Hansen

Preface

Whether you work in aerospace or on Wall Street, whether you are a safety engineer or an IT manager, *Climbing the Corporate Ladder–Safely!* will provide you with career insights and advice found nowhere else. Mark Hansen is well known among safety professionals, but his new book will introduce his frank, results-oriented perspective to a new audience. From his start as a systems engineer with Ford Aerospace to a corporate vice-president in charge of safety, health, and environmental compliance for an oil and gas company, Mark has experienced the ups and downs of corporate life in late twentieth-century America. A keen observer of corporate culture, always first to learn from his own mistakes, and a true believer in the value of continuing education, Mark has written a book that will speak to anyone, regardless of where they are in their career.

In emphasizing his belief in education, Mark has drawn on the works of many other writers for career advice, tempered with his own experience. In *Climbing the Corporate Ladder–Safely!* Mark Hansen doesn't shrink from discussing the major challenges facing each of us if we invest in ourselves to create a meaningful and rewarding–both financially and personally–career. Whether your goal is to move up one rung or all the way to the top, the information and advice Mark provides will help you prepare for that next step.

Career Success

Your Role as the Corporate SH&E Professional

Well, here you are. You've picked up this book because you (1) want to enrich your career; (2) want to find the elevator to success in your company (or the next company); (3) have managed to find the elevator but can't seem to find the "up" button; or (4) have found a button but you don't know whether it will take you up or down. With this book, hard work, doing the right things, a modicum of good fortune, and being in the right place, you could find yourself in the role of corporate safety, health & environmental (SH&E) professional. You want to have a pivotal role and impact on the organization and employees it is your goal to protect. This book will help you find the proverbial elevator and how to hit the "up" button and get to the corner office.

First of all, you need to know that there are many types of organizations in which you may potentially work. There are several different reporting schemes: you can be centralized, decentralized, matrixed, report to legal, Human Resources, the president or the CEO. All of these schemes have their inherent advantages and disadvantages.

A centralized reporting scheme is where the SH&E professional is chiefly located at the headquarters and travels out to the work locations and typically reports to the Corporate SH&E position with a dotted line to operations. A decentralized reporting scheme is where the SH&E professionals are co-located at the remote work locations and typically report to the local operations management with a dotted line to corporate. Either of these can work; it all depends on the people in each of the operations positions who either allow the SH&E function to succeed or fail. Strong SH&E leadership in the field often clamors for a dotted line to operations and a solid line to SH&E. The converse is often true of weak SH&E leadership.

Figure 1. Centralized vs. decentralized and direct vs. indirect reporting to SH&E

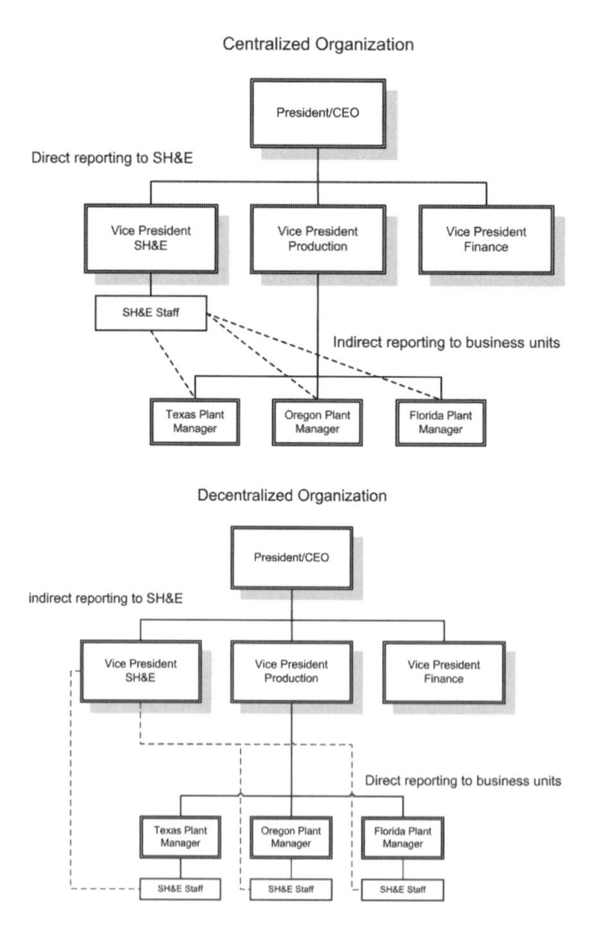

Now the touchy issue of reporting. Again, it all depends on the people in key positions. But there are also some key company idiosyncrasies with each scheme. Reporting to legal can be good—but is often not so good. If the Corporate Counsel has a macro-management viewpoint of doing things right, you will get to do your job. If the Corporate Counsel is a micro-manager, it will present an uphill battle for you to regularly convince your boss from a legal viewpoint, irrespective of the valid SH&E needs. As a result, legal often ties the hands of SH&E, inhibiting you from doing just about anything. For example, implementing a contractor safety program often meets with opposition from legal due to the false assumption that it negates the contracts already in place. You have to explain that OSHA's Multi-Employer Worksite standard requires you to manage contractors.

Reporting to Human Resources (HR) can be good or bad, depending on the person you report to. Most HR people have at least some understanding of SH&E. Some manage by the HR paradigm of "don't upset the apple cart," inhibiting you from doing the job you were hired to do. For example, implementing a drug and alcohol program to meet Department of Transportation (DOT) requirements often meets with opposition due to the impingement on employees. Further, you must explain that the conduct of testing must be kept in close confidence or the word gets out to the field and the desired results are often negatively impacted.

The last and best situation is for the SH&E professional to report to the CEO. There is an onus on the SH&E professional to have your act together or you won't last long. If you do, you only have to sell your concepts to one person, and the CEO will tell the numerous vice presidents what needs to be done. However, if you haven't included the vice presidents in your plans, you may be in line for political execution.

Frequent clear and concise communication is the key to your success. This should occur before you implement programs. It is best that your peers do not discover a program is being implemented after the fact. Using a "no surprises" philosophy will ensure that your peers are kept in the loop of your plans. For example, a monthly report that briefly communicates status and plans is a great tool for keeping your peers informed of your activities, initiatives, and concerns. There is a balance here also. If your CEO is strong, all goes well. If your CEO is weak, then your job gets more complicated. What I mean by "weak" is that the CEO has difficulty making decisions and doesn't want to impinge on your peers. As a result, when there is kick back from the field the CEO often caves in to their demands, leaving you in the lurch. And if your CEO is a micromanager, then look out for headaches. The result is that you need to cover all the bases and have all the requisite data to support your position. Benchmarking can help but often the financial approach is the most compelling.

The best of both worlds is a strong CEO who is not a micro-manager. What I mean by this is a CEO who assumes he or she has hired a person who can make decisions so that the CEO doesn't have to worry about SH&E and getting things done.

Regardless of where you are placed in the organization, you need to have some kind of focus of what you plan to do and how you plan to do it. You need something at the 50,000-foot level with some details. Crafting a vision statement and a mission statement with a few objectives will allow you to have that focus. Management is usually pleased to see that you are thinking strategically rather than just tactically.

If the company already has a vision statement and a mission statement, then take the time to ponder how SH&E fits in and then craft your version. This will demonstrate how SH&E integrates as seamlessly as possible and shadows business objectives. This is usually not a hard sell and management is usually quite open for you to bound the area and focus on how you will benefit the organization.

Writing a Mission Statement

Once you figure all this out, you still may be a little vague on your role and your vision, mission, and objectives. Your first order of business is to articulate and capture these in writing. You need to remember that "SH&E" for all intents and purposes is a verb and the company you work for is the noun. Here is an example to go by.

Vision

To safely <do whatever your company does> for our customers.

Mission

To develop and maintain a proactive culture that reinforces SH&E awareness.

Objectives

1. Establish and maintain relationships with management, field personnel, and contractors.
2. Determine the SH&E needs of management, field personnel, and contractors.
3. Develop viable and usable tools and methods to address SH&E issues.
4. Proactively address SH&E needs through regularly-scheduled visits to worksites.
5. Reactively address SH&E needs through unscheduled visits to worksites.

Tools

Once you have determined objectives, you need to determine what method you will use to accomplish them. These tools are the things you will do to make sure the objectives are accomplished.

Objective #1. Establish and maintain relationships with management, field personnel, and contractors.

Management:

- Attend regularly scheduled management meetings to keep informed of operations activities to proactively address SH&E issues.
- Informally visit with management to maintain good working relationships.
- Participate in new business ventures and provide value-added products and services.

Field Personnel:

- Communicate through regularly scheduled and unscheduled meetings with field personnel.

Contractors:

- Communicate through regularly scheduled and unscheduled meetings with field personnel.

Objective #2. Determine the SH&E needs of management, field personnel, and contractors.

- Conduct needs assessments for management, field personnel, and contractors to determine the areas where assistance is needed to address SH&E issues.
- Conduct hazard assessments to determine the SH&E gaps.

Objective #3. Develop viable and usable tools and methods to address SH&E issues.

- Management systems
- Engineering design approaches
- Procedures
- Audit/inspection checklists
- Training
- Mentoring

Objective #4. Proactively address SH&E needs through regularly scheduled visits to worksites.

- Develop an annual schedule to visit field locations to address management systems, engineering design approaches, procedures, audit/inspection checklists, and training.

Objective #5. Reactively address SH&E needs through unscheduled visits to worksites.

Determine the need to visit worksites based on the probability and severity of SH&E issues in the field:

- High potential accidents and near misses

- New or existing exposures that are imminently dangerous to life and health
- Serious issues the field personnel need to have addressed
- Other viable SH&E issues that are responsive to the needs of field personnel

That's it. It's that simple. It doesn't need to be fluffy and "frou-frou." The mission statement doesn't have to unite divergent cultures like "We Are The World" or solve world peace. Too much pontificating can detract from the focus. This is the basis for what you hope that your organization will do to positively impact the company. You can go as far as putting it on the back of the business card so employees can flip it over and read it in times of angst and frustration.

Why a Corporate SH&E Policy?

This is the beginning of the buy-in from management. This tells employees, customers, and vendors how you do business from an SH&E perspective. It demonstrates that SH&E is a value rather than just a goal. It is usually signed by the president or CEO. So, the next thing you need to do is develop a corporate SH&E policy. In most cases your company already has one, or at least an environmental policy. If your company has a similar one, then just incorporate the SH&E component into it. If not, develop your own. Use your company's core values to develop it or sit down with your management. See Figure 2 as an example.

Figure 2. Example of safety & health and environmental policies

Safety & Health Policy

At our company we are committed to conducting business safely and continually improving our policies and practices for the future. Our aim is to create a workplace culture to integrate safety into all activities, every day. We strive to maintain our corporate goal of an accident-free workplace and will continue our pursuit of excellence in health and safety through the following practices.

Safety Management

We will:

- Conduct our operations in a manner that protects the safety and health of our employees and the public.
- Require employees from all levels of the organization to participate in our health and safety program and both individually and collectively take responsibility to work safely.
- Provide our employees with the required job-related training and safety-related education.
- Seek compliance with all applicable legal and regulatory requirements.
- Investigate incidents and accidents to determine root cause.
- Ensure contractors that work for our company are committed to conducting all business operations safely and in compliance with all applicable laws and regulations.

Continuous Improvement

We will:

- Employ regular audits to enhance successful accident prevention programs as well as to identify, if possible, areas for further improvement.
- Take measures to minimize or eliminate all identified hazards, company-wide.
- Maintain safety statistics for both employees and contractors to track improvement.
- Annually review the overall Health and Safety Program to ensure its ongoing effectiveness.

Figure 2. Example of safety & health and environmental policies (Cont'd)

Communication

We will:

- Ensure all workers, whether contractors or employees, are aware of their right to refuse work that they determine to be unsafe.
- Inform employees and contractors of potential safety hazards on a continual basis.
- Encourage all workers to report immediately and, where appropriate, remediate any unsafe work conditions or activities.
- Openly communicate hazards and emergency response plans throughout the company to affected stakeholders.
- Conduct general safety meetings and job-specific safety meetings as required.

Environmental Policy

At our company we are committed to protecting the environment in which we operate and take pride in conducting our business in a safe and responsible manner. We recognize and accept our responsibility as producers to develop resources with an awareness of the environmental, economic and social needs and expectations of stakeholders. Our commitment is embodied in the following statement of the company's environmental policy.

Environmental Management

We will:

- Integrate environmental integrity, social equity and economic viability into our business processes.
- Provide staff with the resources to make environmentally sound decisions.
- Comply fully with environmental legislation and regulations.
- Carefully manage our use of natural resources and improve energy efficiency.
- Assess the environmental sensitivity of lands, identify impacts and propose mitigation, where appropriate.
- Conduct our operations in a manner intended to prevent pollution, conserve resources, and deal responsibly with past environmental issues.

Figure 2. Example of safety & health and environmental policies (Cont'd)

- Minimize our overall land disturbance on new developments.
- Ensure corporate preparedness with an effective emergency response program.

Continuous Improvement
We will:

- Promote innovative thinking in the development and implementation of new ideas relating to environmental integrity.
- Measure our performance using comprehensive audits.
- Establish environmental targets and objectives to improve our performance.

Communication
We will:

- Respond to the concerns and views of stakeholders in a timely and open fashion.
- Engage interested parties, when necessary, to discuss our business operations and their relationship to affected communities and the environment.
- Provide clear and candid environmental information about our products, services and operations to customers, employees, government agencies and the public, as appropriate.

Now that you have direction, you need to clarify your role. It is equally as a performer, an advisor to management, and an information resource to employees.

As a performer, you have been instrumental in being the catalyst to change the culture, reduce the incident rates, or implement new programs. You have attained positive results, both from your programs and from field personnel, as to your impact on the company. You are an advisor to management regarding SH&E requirements and methods to implement programs. You are an SH&E resource with workable solutions.

Alternatively, your role is not to get assigned all the corrective actions from incident reports. This does not insulate you from doing everything, it just means that your primary role is advisor. Your role is to help provide solutions to management as they provide a safe workplace.

To assist you with delineating SH&E's role in the organization, ASSE's "Scope and Function" is provided in Figure 3. Where this primarily addresses safety, it can easily be expanded to SH&E. Whenever "safety professional" is used in the text, "SH&E professional" can be substituted.

Management must have primary responsibility for SH&E. Management and employees must work together to attain excellence in SH&E performance.

SH&E professionals have wide-ranging responsibilities in performing their professional functions, the major elements of which are identified in Figure 3. Safety professionals must remain technically current in loss prevention techniques, as well as legislative and regulatory requirements and SH&E consensus standards. Functional activities of SH&E professionals include both risk assessment and control of hazards.

A central concept of the SH&E profession is that management remains responsible for implementation of safety measures to control losses. SH&E professionals serve in an advisory capacity to the organization and provide guidance and technical expertise, within the parameters of the profession's ethics, to assist in meeting these responsibilities.

Accordingly, the following are the principal responsibilities for the SH&E professional:

- Assist corporate and line management in assessing the effectiveness of the unit SH&E programs and providing recommendations for maintaining/improving their effectiveness.
- Provide guidance to line management and employees so that they understand the SH&E programs and how to implement them.

Figure 3. Scope and Function of a Safety Professional

Scope of a Safety Professional

To perform their professional functions, safety professionals must have education, training and experience in a common body of knowledge. Safety professionals need to have a fundamental knowledge of physics, chemistry, biology, physiology, statistics, mathematics, computer science, engineering mechanics, industrial processes, business, communication and psychology. Professional safety studies may include industrial hygiene and toxicology, design of engineering hazard controls, fire protection, ergonomics, system and process safety program management, accident investigation and analysis, product safety, construction safety , education and training methods, measurement of safety performance, human behavior, environmental safety , and safety , health and environmental laws, regulations and standards. Many safety professionals have backgrounds or advanced study in other disciplines, such as management and business administration, engineering, education, physical and social sciences and other fields. Others have advanced study in safety . This extends their expertise beyond the basics of the safety profession.

Functions of a Safety Professional

The major areas relating to the protection of people, property and the environment are:

A. Anticipate, identify and evaluate hazardous conditions and practices.

This function involves:

1. Developing methods for:
   - Anticipating and predicting hazards from experience, historical data and other information sources.
   - Identifying and recognizing hazards in existing or future systems, equipment, products, software, facilities, processes, operations and procedures during their expected life.
   - Evaluating and assessing the probability and severity of loss events and accidents which may result from actual or potential hazards.
2. Applying these methods and conducting hazard analyses and interpreting results.

Figure 3. Scope and Function of a Safety Professional (Cont'd)

3. Reviewing, with the assistance of specialists where needed, entire systems, processes, and operations for failure modes, causes and effects of the entire system, process or operation and any subsystem or components due to:
   - System, subsystem, or component failures.
   - Human error.
   - Incomplete or faulty decision making, judgments or administrative actions.
   - Weaknesses in proposed or existing policies, directives, objectives or practices.
4. Reviewing, compiling, analyzing and interpreting data from accident and loss event reports and other sources regarding injuries, illnesses, property damage, environmental effects or public impacts to :
   - Identify causes, trends and relationships.
   - Ensure completeness, accuracy and validity of required information.
   - Evaluate the effectiveness of classification schemes and data collection methods.
   - Initiate investigations.
5. Providing advice and counsel about compliance with safety , laws, codes, regulations and standards.
6. Conducting research studies of existing or potential safety problems and issues.
7. Determining the need for surveys and appraisals that help identify conditions or practices affecting safety including those which require the services of specialists, such as physicians, health physicists, industrial hygienists, fire protection engineers, design and process engineers, ergonomists, risk managers, environmental professionals, psychologists and others.
8. Assessing environments, tasks and other elements to ensure that physiological and psychological capabilities, capacities and limits of humans are not exceeded.

Figure 3. Scope and Function of a Safety Professional (Cont'd)

B. Develop hazard control methods, procedures and programs.

This function involves:

1. Formulating and prescribing engineering or administrative controls, preferably before exposures, accidents, and loss events occur, to :
   - eliminate hazards and causes of exposures, accidents and loss events.
   - reduce the probability or severity of injuries, illnesses, losses or environmental damage from potential exposures, accidents, and loss events when hazards cannot be eliminated.
2. Developing methods which integrate safety performance into the goals, operations and productivity of organizations and their management and into systems, processes, operations or their components.
3. Safety policies, procedures, codes and standards for integration into operational policies of organizations, unit operations, purchasing and contracting.
4. Consulting with and advising individual and participating on teams
   - engaged in planning, design, development and installation or implementation of systems or programs involving hazard controls.
   - engaged in planning, design, development, fabrication, testing, packaging and distribution of products or services regarding safety requirements and application of SH&E principles which will maximize product safety.
5. Advising and assisting human resources specialists when applying hazard analysis results or dealing with the capabilities and limitations of personnel.
6. Staying current with technological developments, laws, regulations, standards, codes, products, methods and practices related to hazard controls.

C. Implement, administer and advise others on hazard controls and hazard control programs.

This function involves:

1. Preparing reports which communicate valid and comprehensive hazard controls which are based on analysis and interpretation of accident exposure, loss event and other data.

Figure 3. Scope and Function of a Safety Professional (Cont'd)

2. Using written and graphic materials, presentations and other communication media to recommend hazard controls and hazard control policies, procedures and programs to decision making personnel.
3. Directing or assisting in planning and developing educational and training materials or courses. Conducting or assisting with courses related to designs, policies, procedures and programs involving hazard recognition and control.
4. Advising others about hazards, hazard controls, relative risk and related safety matters when they are communicating with the media, community and public.
5. Managing and implementing hazard controls and hazard control programs which are within the duties of the individual's professional safety position.

D. Measure, audit and evaluate the effectiveness of hazard controls and hazard control programs.

This function involves:

1. Establishing and implementing techniques, which involve risk analysis, cost, cost-benefit analysis, work sampling, loss rate and similar methodologies, for periodic and systematic evaluation of hazard control and hazard control program effectiveness.
2. Developing methods to evaluate the costs and effectiveness of hazard controls and programs and measure the contribution of components of systems, organizations, processes and operations toward the overall effectiveness.
3. Providing results of evaluation assessments, including recommended adjustments and changes to hazard controls or hazard control programs, to individuals or organizations responsible for their management and implementation.
4. Directing, developing, or helping to develop management accountability and audit programs which assess safety performance of entire systems, organizations, processes and operations or their components and involve both deterrents and incentives.

Figure 3. Scope and Function of a Safety Professional (Cont'd)

Because safety is an element in all human endeavors, safety professionals perform their functions in a variety of contexts in both the public and private sectors, often employing specialized knowledge and skills. Typical settings are manufacturing, insurance, risk management, government, education, consulting, construction, healthcare, engineering and design, waste management, petroleum, facilities management, retail, transportation and utilities. Within these contexts, safety professionals must adapt their functions to fit the mission, operations and climate of their employer.

Not only must safety professionals acquire the knowledge and skills to perform their functions effectively in their employment context, through continuing education and training they stay current with new technologies, changes in laws and regulations, and changes in the workforce, workplace and world business, political and social climate.

As part of their positions, safety professionals must plan for and manage resources and funds related to their functions. They may be responsible for supervising a diverse staff of professionals.

By acquiring the knowledge and skills of the profession, developing the mind set and wisdom to act responsibly in the employment context, and keeping up with changes that affect the safety profession, the safety professional is able to perform required safety professional functions with confidence, competence and respected authority.

- Assist line management with the identification and evaluation of high-risk hazards and development of measures for their control.
- Provide staff SH&E engineering services.
- Maintain a working knowledge and maintain federal, state and local relationships with appropriate regulatory agencies.
- Develop and maintain a high level of professional competence in the SH&E field.

- Represent the organization in SH&E interests outside their operations within the community and by active participation in professional societies and specific trade associations and groups.

So, there you have it. You have your vision, mission, and objectives. You have your signed policy statement. You know where you fit in the organization and your roles and responsibilities. You are now in a position to positively impact the SH&E performance of your organization. You have the framework to play a vital role and have a significant impact on your organization's SH&E programs.

Education and Continuing Education

A generation ago, SH&E professionals either fell into the SH&E profession because they got hurt or they all had white hair because they had been around the block enough to where they had seen just about everything. Some had degrees, others didn't. That's how SH&E was back then.

I can recall just a few years ago when mentoring a fellow professional to get his degree, he said, "I have twenty years' experience, what will a degree do for me now?" The truth of the matter is that a degree may not help you in this job, but it will get you the next one.

That was then, this is now. SH&E is a different animal today. Most professionals have a degree, some have two and some even have three. There is a minority of professionals without degrees that is growing smaller every day. Today, not having a degree—not just any degree but a SH&E-related degree—is all but required. If you don't believe it, just check out the want ads. Today, a degree is the journeyman's license for SH&E.

A word of caution regarding degrees: too many can be too much. For example, if you currently have a BS degree and you work in a plant environment, that should be sufficient. Some plant managers might be intimidated if you have more degrees than they do. That being said, a Ph.D. may be of more value for consultants, corporate-level positions, academia, or other research-based institution (such as

NIOSH). This is where the degree lends technical credibility to whoever you are contracting to for work.

Education is important, not only for passing the hurdle to get the job but also to provide the basic knowledge to do your job competently and professionally. In the past, SH&E professionals learned how to do their job with on-the-job training. As a result, they only learned what they needed to know to get the job done. Myopic and with blinders on, they only experienced 10 degrees of the full 360 degrees of SH&E. Plus, the quality of what you learned is only as good as the instructor. This is particularly true for SH&E professionals who enter the profession mid-career and only know a particular facility and a narrowly scoped application. If you learned a faulty method, you perpetuated the mistake to the next person. Without the full flavor of understanding the big picture of the 360 degrees of SH&E, you miss quite of a lot of how it fits into the overall scheme of SH&E.

When I taught a young professional how to do a fault tree analysis, his comment was, "Wow, now I can compete with you!" I responded, "I taught you everything you know, not everything I know." This is not being arrogant, but rather to make the point that knowing one small piece of the puzzle leaves a great deal out of how it fits into the grand scheme of the systems solution. There are concepts and formulas to learn that apply to disciplines like fire protection, lighting, industrial hygiene, noise engineering, system SH&E, ergonomics, etc. These and other concepts provide the SH&E professional with the requisite tools to do the job for which their employer hired them.

There are a multitude of degrees in SH&E offered today, online or otherwise. There is a partial listing in the appendix. For a current list you can either go to ASSE's website or search the Internet for accredited SH&E programs. This list merely shows that there are SH&E programs all over the country for those who want to learn more and pursue SH&E degrees.

Career Rx: Keep Learning

Change is a way of life in corporate America, so all professionals must continue to learn or risk their career. The choice is life-long learning or getting left behind.

At a major company, a manager hired a candidate with 20 years of experience. But, month after month, the executive was disappointed by the staffer's performance. He just wasn't getting the quantity and quality of work he expected, given the employee's background and experience.

Finally, the executive figured out what was wrong. In reality, the employee didn't really have 20 years of experience. He had one year of experience 20 times. The new hire was not a learner, and didn't seem to have learned much once he got a handle on the basics of his work. He just did the same work over and over.

Non-learners like this person have never been at greater risk of facing career dead-ends, or even of losing their jobs, than they are today, when employees are under constant pressure to keep learning. Companies are, in effect, telling their employees: "We are no longer taking responsibility for your career. You now have to assume that responsibility."

From the employee's viewpoint, this signals the end of job security. However, from management's point of view, the new arrangement says that, as long as companies stay financially healthy, they will provide jobs for employees whose skills match corporate needs. But there are no more blank checks.

That presents a twofold challenge; first, employees must understand the strategic direction of their company, regardless of whether they work in finance, human resources, SH&E, or some other department. Second, employees need to continually evaluate their own skills to ensure that they match the strategic direction and needs of the company.

In this rapidly changing global business environment, a company's only sustainable advantage is based on the knowledge and skills of its employees. Therefore, the only way people can build a progressive career path, whether in the same company or different ones, is by increasing their knowledge and skills. The best way to achieve that is through continuous learning.

A Way of Life

With change a way of life in corporate America, all professionals, not just SH&E professionals, must continue to learn, or risk their career. Today's choice is remaining a life-long learner or getting left behind.

For a do-it-yourself process of learning and self-development to work, three elements are required. These three elements are as follows:

- Continually update and upgrade your skills, as well as your knowledge of your company and industry;
- Hone your interpersonal skills; and
- Sharpen your ability to analyze and solve problems.

It is important of keep your mind open to new possibilities and share what you learn with others. When you do that, you encourage other people to share what they've learned with you. For learners, every day at the office is a learning opportunity. They are able to acquire knowledge every time a question is answered, a new situation arises, or something routine is handled in a different way. Everything they read, hear, and see provides a similar opportunity, as do all meetings, conferences, seminars, and trade shows.

Learners do what gourmets do, they look for and try different restaurants, pay attention to what they eat, and compare each meal with others they've had. Nothing is viewed in isolation. Everything is put in context, in terms of past, present, and future on-the-job consumption.

Dynamic learning-charged managers and staffers also analyze the skills they have and put them to work on the job. These people talk about the business with curiosity, enthusiasm, and interest. They make connections between seemingly unrelated pieces of information and activities, and extract underlying patterns and principles.

Learners are useful performers to watch in action. Take the example of one SH&E manager who met with a plant manager facing a nettlesome problem. When the SH&E manager started to describe how a similar company had handled this problem, the plant manager cut him off by saying, "That has nothing to do with us!"

The SH&E manager persisted and was able to show how the problems were similar even though there were different circumstances. The plant manager's resistance dissolved, and the problem was on its way to being solved. That SH&E manager personified the learner in action. He didn't limit himself to the play-by-play of a case study. Instead, he analyzed past experiences and applied those experiences to new and different situations.

The Learning Bandwagon

Realistically, the call to proactive learning demands careful planning and strenuous effort. Considering the human inclination to "leave well enough alone," along with the widespread feeling of SH&E managers and professionals that they've never worked so hard for so long, it's not surprising that many people don't jump eagerly on the learning bandwagon.

After all, how many people hold post-mortems on solved problems to identify the underlying principles and strategies of the solutions? How many keep files that track what they've learned? How many keep a learning journal?

I attended a conference of SH&E professionals who heard a riveting presentation by an expert who discussed the latest timely issues. When the time for comments came, I asked the crowd how many felt the information they'd heard was invaluable, for them and for the other people in their company. Practically everyone raised their hands. My next question was, "How many of you are buying the proceedings or a tape of the presentation to study and share with others in your company?" Only two out of a hundred people raised their hands.

Continuous learning is one of the critical elements involved in career building in today's companies, which emphasize teaming and the need to work across functions. The desire to continue learning is a differentiating factor among many job candidates today. Take two people trying for the same job or promotion whose objective measurements are more or less equal. That's a tough call. But, if one of the candidates has taken courses or has increased his or her skill set outside a narrowly defined specialty, the choice becomes easier to make. How many of your companies offer tuition assistance? Do you take advantage of it? If you don't, shame on you.

If you're ever interviewing and the interviewer asks you about staying current and continuing education, you want to be able to answer that you are always learning and be able to state the ways you are accomplishing this. Your career may very well depend on it.

Establishing and Sustaining Your "Street Cred"

There are many people in SH&E who think that since they have a degree, they don't need to go out to the field to see what the employees are doing and actually talk to them. Similarly, there are field employees who get promoted to an SH&E position who think that they can now sit in an office and not have to sweat anymore. These people have just reached their level of incompetence. The field could be a remote office, the manufacturing floor, the warehouse–any place where you are likely to sweat. You want the reputation that the employees accuse you of–trying to degrease the facility with the front of your shirt. Because if you don't, when you shun the employees, the employees will shun you.

Why is this important? Little do you know that your "street cred" is based on feedback from the field. I don't care if you are an SH&E professional with a high school diploma or a Ph.D., SH&E is all about building relationships. Building relationships is about getting to know people and developing trust. If you don't like people, especially those in the field, quit SH&E and go do something else. You'll never be successful.

The first thing you do in the field is listen and take notes. Don't talk. We have two ears and one mouth for a reason. There's an old cowboy saying, "Never miss a chance to shut up." There's a reason you can't figure out what others are thinking about—you are talking. When you do talk, ask them questions that are important to them. Usually, the most important topic employees want to talk about is themselves. Ask twenty questions. Get to know them.

Secondly, you really need to know your stuff. That is, you need to understand SH&E and whatever application you are working with, a manufacturing process, a chemical process, organic chemistry, nanotechnology, or whatever. Learn the process. In the same manner, you need to know how SH&E applies and where it applies. This builds credibility with employees, showing that you can really help them rather than just wasting their time developing esoteric procedures.

Thirdly, include employees in the process. Rather than making their procedures fit into the regulatory requirements, make the procedures meet the regulatory requirements. Start from their perspective rather than the regulatory one. Help them make it fit rather than telling them that they need to make it fit. After all, you are the SH&E pro, not them. It doesn't have to be perfect, it just needs to work. This will increase your credibility. And remember—Noah built the ark, engineers built the Titanic.

Fourth, once you build relationships, keep on building them. That means continual communication. This communication is both technical and personal. Ask about what their kids are doing, what trials they are going through, and oh yes, the job. Nothing beats old-fashioned caring. After all, people don't care how much you know until they know how much you care. It is so true. Many people believe that you can build a relationship with email. Yeah, right. You can sustain a relationship for a short period but you cannot build one by email. Listen closely: email is primarily for sending documents and sound bite communications. Don't even think you've built a relationship on email.

Remember you have never arrived. That is, be humble about yourself. Humility is one of those things that once you know you got it, you just lost it. Never forget where you came from and the employees will rely on you and call you, not only for help, but just to talk. Be there for them. You are never too big for your job. If you get prideful, remember that pride is a disease that makes everyone around you sick. That's what we are there for. We are the closest discipline that actually cares about the employees. Show it.

Building your street cred comes from getting to know the field employees and building relationships and trust. If you do that, the field will call the home office and rave about all the good things you've helped them with. Who knows, they may even buy you lunch on your next road trip to the field. That's street cred.

I'm Too Smart For This Job!

Do you feel that you are under-employed, not using all of your talents and capabilities like you would prefer? Do you feel that you could do your boss' job? Well, join the club. Many experienced SH&E professionals are losing out in the job market, or are hanging on to boring, unchallenging jobs until retirement.

The Real Story

In my discussions with fellow SH&E professionals during the past six years, many have settled for positions with lower pay, status, and skill requirements than they held previously. Some accepted their plight because they couldn't afford to wait any longer for better positions. Others found that the kinds of jobs they'd left behind no longer exist. Still others survived company restructurings, only to find that their jobs had diminished in scope.

These people are underemployed, which can be defined as not being used to the fullest extent possible. The Bureau of Labor Statistics doesn't tally the number of underemployed white-collar professionals, much less SH&E professionals, but industry observers agree that their ranks are reaching new heights. Faced with stiff competition for new jobs, many job candidates lower their standards when considering offers and end up in positions that are less challenging than they expected.

Fear and advancing age also contribute to underemployment's growth. Older professionals typically don't find positions quickly, especially if they aren't up to date with current technologies. If they're

55+, they can have wonderful technical skills but still not get interviews. Newspaper articles about the economy rarely mention that many older candidates are taking jobs below their skill level. Instead, they cite shortages in many specialized fields.

Facing the Situation

There are many ways to define underemployment. But earning less and having a smaller office doesn't necessarily mean the job isn't challenging. Small businesses have more than 60 percent of the jobs, and they usually offer less base salary and more incentives, such as stock options. Because small firms tend to offer terrific opportunities to mid-career candidates, they should be researched thoroughly during any job hunt. And if you're willing to live without many of the traditional status symbols of being an executive, you can broaden your employment horizons without being forced to accept a job you can do in your sleep.

In the old contract, the size of your office mattered. Some people are given a cubicle and feel they aren't recognized, but recognition isn't by perk any more. It's by the value you add. Job content is changing as well. Many professional jobs now require hands-on work that was once considered demeaning.

Adapting to workplace changes requires a flexible attitude. If you think of yourself as a corporate SH&E manager in a medium-sized company, you may find there are few suitable positions. It's different if you define yourself as a SH&E professional who is an advocate for change, and recognize that your skills are transferable.

Thinking It Over

Many people are under-employed because they accept jobs too quickly. If you can afford it, take four to six months off to regroup, especially if you've left your job involuntarily. The stress of being in an unsuitable job or an unhealthy organization is exhausting. Sometimes you have to heal. Then you may be more open to different possibilities.

If a mortgage and other bills can't wait, there are other ways to test the waters, such as seeking a consulting assignment to explore new options and gain experience. Because it's such a competency-based workforce, the more skills you develop, the better.

Temping is another worthwhile alternative. It may feel better than taking a lower-level permanent job, because you think you're in transition. Temporary work has proven to be a great avenue for bringing in income, getting out of the house, and keeping current with your field, and it often leads to other things (sometimes even a permanent position).

Security Suffers

When accepting a less-challenging position, many SH&E professionals are motivated by job security. But like it or not, we're all temporary employees. These days there are no job guarantees. If you're having trouble accepting this new reality, it's likely you've developed a victim mentality.

People vary in their capacity for change. I have seen about four distinct response patterns among professionals faced with career dilemmas:

- Overwhelmed (low comfort with change and low capacity for handling new duties).
- Entrenched (low comfort and high capacity, they cling to past values).
- BSers (high comfort and low capacity, high drive but can't deliver).
- Learners (high comfort and high capacity, they deal positively with change and learn relevant new skills). The learners are the ones who thrive.

Technology, specifically computer literacy, has taken a toll on middle management. Middle managers used to be the custodians of organizational values, like tribal elders. They're not strong enough to hunt, but they were kept around because they told stories and passed messages between the bottom and the top. Then along came the computer, and companies didn't need middle managers to communicate, they could do it with computers and talk directly to customers. So these managers hit the streets in hordes.

Unfortunately, many haven't taken the time to learn the relevant skills. So when they come back, they're underemployed. All you have to do is find the right situation. You need the skills, and if you feel you have skills that your employer doesn't recognize, you have to tell them in ways they can understand.

Getting Out

What if you're already trapped? If you don't feel good about what you're doing, and you can't change it, it's fight or flight. Flight isn't bad. There are jobs out there for managers with a good work history and skill set. Never stop looking. It's better to risk age discrimination than to sit around in a bad situation. And when you network, emphasize your skills and added value. Don't even talk about your age. Talk about indexes of success, such as incident rates, cost savings, and worker's compensation dollars.

Strive to emphasize skills and results, rather than experience. Develop a portfolio of stories, and then find colleagues who will give you references. Bosses and subordinates still count, but it is nice to have alternatives. Just because you've moved down, it doesn't mean you can't move back up. People who were SH&E managers sometimes have to go back to engineering. The astute ones who kept up their technical skills and can sell themselves can climb back. Lots of companies work on moving them back into management.

For others, there is a trend toward holding multiple jobs, or "portfolio careers." People have to put together several of their skill sets to maintain their net worth and lifestyle. They may have several consulting jobs, for instance, which can be positive in the long term. They expand their skills and may eventually return to their former salaries, or higher ones.

Another skill is learning to explain why you're underemployed to interviewers. It's important to have a believable explanation for being in your job. If it's for economic reasons, and the market in your field is very tight, that's fine. But if you're hesitant, people may feel there's more to the story than you're telling them.

Enjoying It

Giving up a lot of the responsibility also means giving up a lot of the headaches. This provides more time for family and other extracurricular activities and less job stress.

The other benefit of underemployment is that there is room to grow. It's better to be involved with an upwardly-mobile company than to try to save companies in tired old industries. People in declining industries sit around and tell you they're having fun, but I don't think a lot of them are. Fun is working with an emerging company in a growing field with a process that has the potential to revolutionize an industry.

Time to Retool

Some underemployed professionals use this time in their lives to retool. This can mean going back to school for another degree or a higher degree in the same or another field. Changing careers entirely is another option. Explore two or three tracks, especially if you're in a declining industry or function. You may choose to do a Myers-Briggs assessment, then go out and explore the tracks through informational interviews, research, and maybe by consulting. Start doing volunteer work in an area you enjoy. You may find a new career that you could explore.

If you can't spare the time or money to look for your ideal job, concentrate on fulfilling your other interests. Even if your job can't change for five years, you can start looking at other fun things now. Hobbies and other interests can keep you going. Start finding activities outside of work that will keep you busy into retirement. You can look forward to enjoying a different part of yourself later.

Career-related volunteer work also can be satisfying. The good news in an improving economy is that you probably can escape a bad situation and try something new without earning the stigma that once accompanied demotions and job-hopping. Bright people can do a lot of different things. I think each of us has three or four careers in us.

So if you think that you are too smart for this job you can find ways to cope, improve yourself to rectify the situation, or just enjoy the change. First think it through and realize that the days of job security are gone. Make the best of your job and take as many challenges that come your way.

Ten Commandments For Career Success

Succeeding in the real world doesn't require being in the right place at the right time, knowing the correct people or even having superior luck. You can achieve professional success without a miracle, as long as you follow some practical tips while launching your career. By following these ten commandments, you will enhance your work habits, philosophies and coping methods for your career.

X. Know thyself

Know what valuable strengths, skills and experiences you can offer employers. Make a list of the technical, managerial, organizational, interpersonal, communication, and computer skills you've gained through your career, continuing education, volunteer work with professional societies, or other types of volunteer work. An accurate assessment of your abilities will help you set realistic job-search goals. If you're unsure how your experience translates, ask trusted colleagues or mentors for assistance.

Also know what fields interest you. Fields in SH&E range from process SH&E to auditing, insurance to transportation, engineering to environmental, among others. If your options seem limited, broaden your horizons by taking on additional tasks related to your interests at work, do consulting on the side, or even teach at a university. Talking to others about your goals, interests, and experience will help you to know yourself better and create a suitable career plan.

Keep in mind, however, that this broadening of your background could easily lead to a career veer into a related field or upper management. For example, you may end up in a corporate SH&E position that ultimately lands you a vice-president slot, at which point you will exit the field of SH&E. Your next stop may be acquiring an MBA at a local university to give you a firm foundation in the business field.

IX. Know thine industry

Know how you might fit into the fields that interest you. Investigate specific occupations by consulting your local library. Such resources include Dun & Bradstreet and Standard & Poor's. Read relevant ads and job descriptions to determine employers' needs and typical salary ranges. Check out the job placement centers at conferences and inquire to the responsibilities, salaries, etc., of the positions in which you have an interest. Determine the positions in your target field that are appropriate at this stage of your career and the competition for opportunities you want.

Also, check out the trade magazines at the library and online. Subscribe to all of the free magazines and read up on the industry. Understand the hot topics, regulatory issues, and who the major players are. If possible, seek to publish in these trade magazines to get visibility to your peer groups in this industry.

VIII. Thou shalt network

Networking is an exchange of information that isn't just for business people. You can network in airplanes, supermarkets, gas stations, jazz clubs, parties, and community, church, or club meetings. Even if you're shy, you can find a style of networking to suit your personality.

Your goal is to get to know professionals in jobs you dream of having and find out how they got there. Start by asking people you know well—trusted colleagues, family members, friends, old college professors—who they know working in your target fields of interest. Call or write these new contacts and ask for a brief, perhaps 20-minute meeting over lunch to learn more about their field, receive input on your job-search strategy, and get names of other people to contact. Using this technique, you may also find out about unadvertised positions. Don't be afraid to ask, they can only say no.

To make a good impression, be interesting and interested. If you pass along career and job information to contacts, chances are they'll reciprocate. Send thank-you notes after the meeting and stay in touch. By keeping in touch you may discover those unadvertised positions and make your career veer sooner than you thought.

It's also wise to join professional organizations in your field. If you are in SH&E, join ASSE. If in industrial hygiene, join AIHA or ACGIH, and so on. These organizations publish newsletters and hold meetings and annual conferences, which are great ways to meet people and exchange information.

VII. Research thine options

Before jumping off into the world of consulting, do it part time. Determine if consulting is for you. It is not the glamour job many think. You have to be ready when the customer wants you to work, and it's usually on the weekends, when they don't want to work. Regardless, think hard before you consider changing careers.

Information creates opportunities, so read trade journals and local newspapers to learn about industry news, trends, and key players. Write to executives profiled in articles about your field. Express interest in the news story and their companies and request an informational interview. You just may have the ideas and skills they seek.

VI. Cultivate a good attitude

When interacting with hiring managers and networking contacts, be positive, resilient, flexible, and professional. Be able to deal with rejection as well as acceptance and keep your ego in check. If you have a chip on your shoulder, are discouraged or feel depressed, find ways to improve your attitude before you schedule interviews. Otherwise, your true feelings will show through despite your best efforts and hurt your professional image. Don't take anything personally. If it is constructive criticism, improve yourself; if it is destructive, you be the judge and discard it.

V. Think like an entrepreneur

Use your imagination and creativity when job hunting. Spend time brainstorming with friends and colleagues. Think broadly about industries and occupations that might need your services, since career opportunities may be available in unexpected places.

Take a lesson from musicians, the quintessential "multi-preneurs." Their careers often involve performing with various groups; tutoring privately; teaching; recording, composing and arranging, and administering arts programs. List the types of work you'd find stimulating and profitable. You may decide on a varied career being a generalist or on a narrowly-scoped field being a specialist. Be careful not to specialize too much where you end up knowing everything about nothing. For example, in system SH&E, if you become an expert in Failure Modes and Effects Analysis (FMEAs), you may discover that you lack the industry experience to be credible or that very few FMEAs are really being performed anyway. The same is true for being a generalist. You can fall into the trap of not knowing anything specific and know nothing about everything. You don't want to become the office "know-it-all."

IV. Have a gimmick

Persuading employers to hire you is easier if you offer something they want that's unique and exceptional. Revise your resume, focusing on the talents, skills, and experience that make you stand out. The key to a good resume is "why me and why not the other person". Write a skills summary or professional profile, a 20-second commercial about yourself that you can use in resumes and cover letters. In it, be crystal clear about what sets you apart from your competition in the field. Try to keep it between 1 ½ to 2 pages as a rule.

III. Respect thine interpersonal skills

Industries are small, relationship-driven worlds, especially in SH&E. Someone you snub today may be the person who won't hire you tomorrow. This includes headhunters, secretaries, and junior professionals. Don't be known as the griper or malcontent. You definitely won't go far if that is what people remember you for. So be a considerate, polite colleague and help create a healthy workplace community. Your thoughtfulness, optimism, and enthusiasm will be remembered and returned.

Even if you've suffered disappointments, frustrations, and emotionally difficult periods, avoid inflicting your personal difficulties on others. Get whatever help you need so you can be at ease with yourself and others. This will help you benefit from and appreciate your professional affiliations. Also, don't slam your previous employer.

II. Set short- and long-term objectives

Goals are dreams with deadlines. Picture the job you want: this vision should keep you motivated during your job search. Then establish short-term goals, for the week, month, and year to keep you progressing toward your ultimate objective. Everything you do should be a measured step in accomplishing your long-term goals. However, don't let your goals strangle you, keep them flexible enough so you can respond to changing markets and the range of opportunities that may come your way.

I. Feed thy soul

What do you find valuable and meaningful in your life? What keeps your spirit alive? Whether its spirituality, family relationships, the outdoors, or a favorite hobby, fulfill these needs regularly. Don't spend all your time, energy, and emotions on your job search or career. Having a balanced life is critical to enjoying and moving forward in your professional career. One of my colleagues climbs mountains. His goal is to climb all the "fourteen-ers" (all mountains over fourteen thousand feet high) in Colorado. Another is deeply involved in the church. Feed your soul and balance your life so that you actually have a life and don't live to work, but rather work to live.

Well, there are my ten commandments for business success. If you can focus on these tips and grow your career, you may find your career blossoming. Grow for the gusto!

Taming Your Workplace Demons

As the manager of SH&E services for Union Texas Petrochemicals in Houston, Texas, I was hired in March of 1997. I truly hoped that this would be my last job change. Alas, in March of 1998, just one year later, ARCO surprisingly purchased Union Texas Petroleum (which included Union Texas Petrochemicals) with a fast-paced closing date of June 30, 1998. Careers were quickly and indelibly changed on that fateful Monday morning. ARCO had no intention of sustaining the

Houston office. Few were expected to transition to ARCO ... Yuck! break out the old resume once more.

Could your seemingly secure position at a long-term employer also be at risk, too? As companies attempt to squeeze more profits out of their operations, downsizing, mergers, realignments, and re-engineerings are common. Faster, more efficient technology and a competitive global market, where cheaper producers spring up overnight, also make forced unemployment a reality of today's job market.

In this environment, the secret to staying employed is to be flexible, confident, and marketable. If you haven't adopted such an attitude, don't delay any longer. In this tumultuous time, you can tame job loss, insecurity, and other workplace monsters by taking these steps.

Confront your demons. You can't refuse to accept that times have changed. Many SH&E professionals cling to the notion that they'll be with their current employer for life. When pink slips arrive, they complain that life dealt them an unfair blow. Ignoring or wishing away a pending job loss will only make a bad situation worse. You must face your frustrations and insecurities, expand your concept of employment, and explore options that you may not yet realize. Ultimately, facing painful events and making courageous choices will help you to learn, grow, and change for the better. You'll become more confident when you realize that employability has nothing to do with a company's benevolence and everything to do with your skills. Having certifications, good degrees, and requisite experience plays a strong role.

Ask pertinent questions. If rumors about layoffs are swirling, ask questions to learn if the information is accurate and about the scope of any pending changes. Decide if your source is reliable, then learn what the company grapevine is saying. Many times the grapevine is more reliable than formal statements. What's driving the changes? Is the company downsizing, re-engineering, or flattening management? Where are pink slips most likely to hit? Will your job be directly or indirectly impacted?

Also learn about long-term issues that might affect your job. Are global forces making inroads? Will new technology threaten jobs? Shareholder value dictates what will ultimately happen. What can you do to remain a valued player? Answering these questions will help you make informed choices. No one has a free ride. We must be able to justify our position and role in a company. The bottom line is simply that you must illustrate your value for a company.

Control your destiny. Realizing that you have the power to take control is the first step to actually doing it. Unfortunately, many SH&E professionals don't realize that they have the ability, to affect what happens in their lives through their own efforts. To take charge of and manage your career, you must realize that you're the "locus of control," the place where change occurs. The key to assuming personal power is giving up the notion of loyalty to one company or that the company is responsible for your destiny. Companies can no longer afford to offer you the security they once did, nor can you afford to offer them unquestioning loyalty. You can no longer depend solely on your employers to keep you at the front end of the state of the art; the fundamental responsibility for your learning and growth rests with you.

See your career as a developmental journey, and seek advice on converting obstacles into opportunities from corporate training consultants and human resources professionals. Continue to grow and learn. Do things that may not necessarily help you in this job, but would surely help you in your next job. Your actions can mean the difference between keeping or losing a current job. Even though you'd probably prefer to allow fate to make choices for you, you must take an active role in reinventing your future. Your path may take unexpected detours, but you'll be in charge, not your employer. These detours could include similar or divergent industries, different roles based on recent experience, including moving up the food chain in the organization of a new company.

Be prepared for the unexpected. Accepting career upheaval means taking advance action to maintain your own and your family's happiness, regardless of what happens to you professionally. This means being prepared in two ways.

First, be ready for whatever might happen professionally. Stay abreast of changes that might cause you to lose your current job. If you're laid off, be ready to search for a new job, or even change careers if necessary. When those headhunters call monthly, be nice to them, even though you may be busy. You never know when you will need them. Maintain a good relationship. I always keep my ear to the ground and refer fellow professionals I know are searching for better jobs to headhunters looking for talent. That way, when I call, they remember me. In today's rapidly changing economy, it's normal to continually reevaluate where you fit within a company and to look for improvement.

Second, review your finances and take steps to remain solvent in case of unemployment. Exercising financial responsibility is critical to gaining peace of mind. To do this, you must control your expenses and increase your savings and investments.

If your entire paycheck is required to maintain your standard of living, consider adopting a less lavish lifestyle. Postpone major purchases, such as a car or furniture, until your job situation is stable. Don't act in a vacuum. Involve family members in budget conferences and let them know that changes may be inevitable.

Accept that you have choices. Even if a downsizing looms, you don't have to let destiny take its course. Instead, you can make choices that will improve your outlook. Have you wondered why some employees keep their jobs following layoffs, while others don't? Be the kind of employee who will be asked to stay. Cultivate and offer skills and services that are difficult to replace. One desirable skill is finding innovative solutions to tough problems. Corporations need managers with creative solutions to problems. Volunteer for activities, such as special projects or charitable events, that highlight your ability to effectively lead or participate in teams. Be positive and compassionate, not negative or destructive, and others will gravitate to you. The valued employee must wear different hats, assume multiple job roles, and be increasingly flexible.

Develop support systems. In crisis, the value of encouragement and positive reinforcement from family and friends can't be overstated.

Spending quality time with such supporters needn't be expensive and can take the edge off your worries at work. Participate in activities that allow you to vent frustration or keep you so busy that you don't have time to obsess about work. Develop hobbies, join clubs, exercise, read, or do other productive things that prevent you from worrying. Exercise regularly, since it will help you maintain energy, manage stress, and project a positive image. The goal is to gain balance and perspective so you won't be overwhelmed by turmoil.

Polish and update your skills. Stay abreast of changes in your field, and keep your technical skills current by taking courses at work, community schools, or training centers. Embrace new technologies., Attend conferences and events that allow you to hear new ideas and meet contacts in other industries. If funding is minimal, at least attend the ones in your city or town. It even pays to take a vacation day to broaden your background.

Cultivate contacts by becoming a useful resource. You never know when you may need to ask a favor in return. Visibility is key to making swift job changes. Create a high level of activity; the more people who know about you and what you want, the better your chances of success. Assess your professional strengths and skills and then update your resume. Include as many accomplishments as possible. This will build self-confidence at a time when you need it most.

Stay informed. Unfortunately, I have seen more SH&E professionals bury their heads in the sand when the going gets tough. When the going gets tough is when the tough get going and network in the profession. Know what's going on in your company, industry, the job market, and the economy. Read newspapers, books, and business magazines. Pay attention to economic indicators and identify growing industries and companies.

Who's increasing production, building new plants or facilities, creating new divisions, or improving services? Track what's happening in these companies while maintaining a high profile at your own. Also learn your employer's severance policies. Review health insurance coverage, retirement plans and how annual and long-term incentives are paid. This will help you to know what to expect and ask for if your job is eliminated.

Find a security blanket. Accepting that your job can be eliminated at any time may feel like having a cherished childhood security blanket taken away. Children often find other items to tide them over during the transition, and so can you. Substitute family activities, hobbies, or volunteer work. Helping others is sometimes the best way to feel better about ourselves. Realize that negative feelings, such as wanting to lash out, are temporary. Be dignified and diplomatic. Angry outbursts are as ineffective now as they were in childhood.

Recognize the identity factor. Many SH&E professionals define themselves by their jobs, and they use these roles for identification and direction. The thought of losing your identity may give you the chills, but your job isn't who you are. Believe that you're preparing yourself to be a better person or to have a better role. Believing this will help it to come true.

Explore new options. Be open to new opportunities in your field or unrelated industries. Changing your career path doesn't imply failure. In fact, it's just the opposite. Flexibility, resilience, and the ability to bounce back from traumatic events are signs of health and self-esteem. Being flexible gives employees an "I-can-do-other-things" attitude and a broader skill base. Flexibility makes employees more marketable and more valuable. Developing contingency plans and evaluating all alternatives and options, such as career change, consulting, small business or buying a franchise can also help add to your flexibility. By increasing your options, you'll improve your prospects of landing a job more quickly. If you want to catch many fish, you need to get a lot of hooks in the water.

Have best- and worst-case scenarios. I am an optimist. I look at the up side. My wife is a pessimist (she refers to herself as a realist). Together we balance each other. I come up with the best-case scenario. My wife comes up with the worst-case scenario. Defuse your fear of the future by identifying the worst thing that might happen to you and your family, then set the thought aside. Now think of the best things that could happen and concentrate on making this potential come true. A career change isn't a prescription for disaster. Be someone who achieves dreams. After all, why shouldn't you be the one who survives in style? You must be positive and persistent to realize your career dreams. People like to be around happy, upbeat, fun-to-be-with people.

View this period as a learning opportunity. Success can make us complacent or oblivious to the rest of the world, while a job crisis turns the world upside down and jolts us back to reality. You can regain your balance by viewing career transition, learning, and growth as the norm. View each phase of your career as a bridge to a new opportunity and take advantage of this chance to learn. You won't be plucked out of obscurity for a great position. You must do the hard work. Don't give up or expect employers to pursue you, even if they do. Retain control and follow up.

You don't have to look far to find people who are worse off than you. Their plights may involve far more serious tragedies than a temporary career setback, and they desperately need encouragement and help. Instead of automatically seeking a demanding position that requires travel and long hours, reexamine your priorities and decide if that's what you want. You may discover a deep inner strength and faith that you had never put to the test before. Consider the example you want to set for your family. This is a chance to show that you can solve seemingly unmanageable problems through commitment, responsibility and a solid work ethic. You'll become a better person along the way. By conquering job monsters, you'll gain a new sense of confidence in yourself and your employability.

How to Have It All

Making sure you enjoy a high quality of life seems like a straightforward process. All you have to do is simply figure out what makes you happy and incorporate those activities into your daily life.

But when you factor your career and financial needs into the mix, this simple tenet can become extremely complex. Can you progress in your career and have the life you want, or are the two mutually exclusive? Nowadays, more people want both, but they don't always get it.

Many professionals change jobs specifically to make more time for their families or personal interests. Someone looking for a career change is usually seeking a change in an aspect of the quality of their life.

But if making more money is your number one priority, your career decisions won't necessarily produce a better quality lifestyle. While you may earn more money, you may have less time to enjoy it. Many SH&E professionals I speak with today say they want more balance in their lives, but are afraid to make necessary changes. For them, a career change is like a whimsical dream. It has little merit in this fast-moving, quickly changing world. A lot of SH&E professionals talk about quality of life without doing something about it. There's the fear of sacrificing income, status, and security.

Working to Live

Today, many SH&E professionals live to work, rather than working to live. I live by the latter. Instead of judging a position solely on its salary or title, I factor in quality-of-life issues. When I focus on these concerns, the jobs that I consider ensure that I will be able to enjoy life.

Priorities seem to be changing. I have personally witnessed SH&E professionals whose priorities focus more on the quality of life over salary. Similarly, some are willing to work fewer hours for more personal time or time with family. For example, some are tending to aging parents in the twilight of their lives or assisting them during terminal diseases.

As more working parents step off the career fast track and seek greater fulfillment in their personal lives, employers will need to respond with a more flexible work environment. The quest for quality of life isn't limited to aging baby boomers who want to redefine their professional goals. Baby boomers' kids feel the same way.

I have found that throwing money at a job applicant fresh out of school only goes so far. Their quality of life is a major issue. For example, a young SH&E professional is often required to travel to different geographical locations. This is an impediment for life after work.

They are only using their homes or apartments as a weekend residence. When you factor in pets and family, the issue gets bigger. Being flexible to where they reside is almost a must to consider. For example, if their family is a six-hour drive from their work and they are on the road weekly, the company could allow them to live closer to family and friends. Another issue I have found to be a driver, is allowing for personal and professional development. Going to conferences and chapter meetings allows for networking and growth. When that is incorporated into the equation, the whole offer is more appealing.

The Right Choices

No matter how much money you make, you'll never be happy if you don't have enough time to enjoy life. To incorporate a meaningful lifestyle into your career decisions, consider these steps:

Think through your career plans. Consider how a job change or relocation will affect others in your family. For some employees, moving isn't a realistic option because of family considerations; close ties with friends and other factors also play a role. You need to think really hard before making a drastic change. People need to do their homework. No one should walk blindly into a job search.

Take small steps. People can't make a change in one step. It's unrealistic. For professionals who are uncertain about their careers, put together a five-year plan in small discrete steps and revisit it frequently. Formulate attainable goals and realistic steps for achieving each milestone.

If you want to change industries, seek transitional positions and companies that allow you to gain experience that can bridge the gap between your current and projected position or company. If a part-time job proves successful, it may transition you to your career veer company.

This strategy worked for me when I changed from aerospace to the petrochemical industry. I started consulting on the side and teaching at two local universities in SH&E and ergonomics, both graduate and undergraduate courses. The teaching and consulting allowed me to diversify out of the pigeonholed aerospace label. I first went to a Department of Energy contractor that also had commercial business. I soon transitioned to the specialty chemical industry where I was able to make an impact. I originally thought this would be my last career change. The lure of consulting caught my attention. After deciding that consulting was not for me, the best job I have ever had came along. Here I am and I am here to stay, as long as my employer wants me to stay.

Most professionals give up because they fail to change their lives in a single step. As you can see, transitional jobs allow small career changes to take place without suffering a dramatic loss of income or status. Don't just sit around and complain about your current job. Zip your lip and do something about it. I encourage SH&E professionals dissatisfied with their employers to take more responsibility for their lives and happiness instead of making sudden dramatic changes. If you want a better job, take action.

Dream a little dream. Follow your dreams, don't just have them. This isn't always easy. It takes courage and involves measured risk-taking, learning, growing, and maturing. Be sure to differentiate fantasy from reality so you won't pursue impractical or unattainable dreams.

Find your gifts. Everybody has natural gifts, it's just a matter of discovering them. To find yours, there are at least two approaches.

- Ask others to point out your strengths. Some of your talents may come so naturally that you take them for granted. Consult with colleagues you trust, close friends, family members or co-worker to find out their views of your abilities.
- Review your past successes. In each case, why did you succeed? What was the common denominator? Allow these clues to lead you to your ideal job.

If you still can't connect with the right career, ask yourself these questions:

- Is this the right goal?
- Am I taking the right steps?
- Am I being too ambitious?

Review your answers to find the variables you can influence, then focus on what you can change and control. As Oprah Winfrey and many others have noted: You can have it all, you just can't have it all at once. And lighten up! It helps to have a sense of fun to sustain you. If you don't have fun at what you are doing you will never be happy at what you are doing.

Bigger Doesn't Always Mean Better When It Comes to Company Size

Even though I hail from Texas, whose residents seem to never tire of saying, "Everything is bigger in Texas," bigger isn't always better, especially when it comes to company size. In a smaller company there are by definition fewer people. Employees wear many hats and cover many diverse disciplines. For example, in SH&E you may also be tasked to cover risk management, insurance, a little bit of subcontract management, and Responsible Care™ (the chemical industry's voluntary SH&E initiative), in addition to environmental and health. By doing many different SH&E-related activities you can fulfill more of your potential as a professional while at the same time increasing your job satisfaction, and staying happy in the long run.

Conversely, in a large company your whole job description may be to implement only one of the seven codes of Responsible Care™. You may have some job security for that particular area, but if, God forbid, you get laid off, your skills are quite narrow.

I once worked for a huge company of over 30,000 employees. Because of the bureaucracy, red tape, politics, the "not invented here" syndrome, micro-management, and the fear of failure, getting anything done moved speedily at a snail's pace. Working for a smaller company there is fewer of these barriers. Things may not happen at lightning speed, but they happen at speeds several orders of magnitude faster than at larger companies.

Having left a big company for a smaller, more agile entrepreneurial company allows me to make a greater impact on my company, my colleagues, and culture, while charting my own course. Although my transition took seven years rather than my original vision of five, and

I had to make some sacrifices along the way, I do not long for a return to a large company. I'm having way too much fun.

Some of the people I speak to who are contemplating career moves often hear that working for a small company can be frustrating, require more work, or be financially unrewarding. But many negative comments about smaller companies are truly myths based on misconceptions and stereotypes or perhaps even bad experiences from colleagues who just couldn't make the transition. As there is nothing new under the sun, these myths aren't new. Professionals desiring to move to smaller companies for whatever reason (such as, being laid off, re-engineered, downsized, or merged out, or because of internal politics) may wish to take note of these myths before changing jobs.

What follows are some common myths about small companies.

Myth: Small companies don't employ as many people as large companies.

In the US, small companies employ far more people than large companies. I'm talking here of small companies on the order of 10–15 employees. These companies attract relatively little publicity so their impact on the economy is discounted. However, as a group, small companies form a powerful, dynamic economic community which tremendously affects domestic and global growth.

Myth: Salaries are lower and benefits are less generous at small companies.

While many firms, especially slow-growing, family-owned operations, often pay less, other small companies offer equal or greater compensation packages than their larger counterparts. These companies are often started because of a founder's desire to earn greater income. While going it alone may be riskier than working for someone else, those who succeed in the right niche reap handsome rewards. Let it be known, there is definitely more risk. You will have to determine how much risk you want to assume.

Start-up companies funded by venture capitalists may pay well because they need to produce immediate results for investors. Often, a young organization may bolster employees' cash compensation with generous stock options. If the corporation goes public or finds a buyer, the options can literally be worth millions.

Also, smaller companies may actually have better benefits (e.g., medical, dental, vision, 401K, and pension plans). It is ironic that smaller companies realize that people stick around more for the benefits than anything else, especially if they have children. Large companies see this as a loss expenditure, smaller companies see this as an investment in the employees.

Myth: Jobs are less secure at small companies.

Not any more. The golden days of employment for life are long gone at large companies. In fact, many small companies are founded by executives who repeatedly have been downsized or re-engineering and want more control over their careers. They trust their abilities and decisions more than those of a large company. Employees at smaller, well-meaning companies may feel more secure about their jobs because of a personal relationship with the owner. Smaller company owners aren't pressured by stockholders' expectations for profits, and will move heaven and earth to make payrolls. At these firms, layoffs typically are used only as a last resort.

Myth: Small companies offer fewer opportunities for advancement.

At family-owned, sole-practitioner, ill-managed, or stagnant small companies, this may be true. However, fast-growing and profitable small companies must continually add employees and managers to keep pace with demand for their products and services. Employees in these companies wear many hats simultaneously, so on-the-job training, continual learning, and increasing responsibility are required.

Myth: Large companies are unlikely to do business with small companies.

Downsizing and re-organizations at large companies have created opportunities for smaller companies to provide products and services that were once performed in-house. How many professionals do you know who have returned to former employers as independent contractors after being laid off? Can you say outsourcing?

Large companies are more interested in working with smaller niche market companies because they often receive higher quality products and services for a reasonable price than they do from larger competitors. Many larger companies and government agencies also must meet internal affirmative action goals that require them to contract with minority or woman-owned firms. This creates even more opportunities for small businesses.

Myth: Small companies use the same hiring criteria as large organizations.

Most large corporations rely on education and experience requirements to screen new hires or promote existing employees. But small-company managers tend to be more open to candidates with different backgrounds. While they may seek certain experience, they consider enthusiasm and aptitude when making hiring decisions. They tend to trust their own instincts and are more willing to hire career changers or less educated professionals who demonstrate an ability to do the job.

Myth: You can't make a big contribution at a small company.

One of the reasons people move from large companies to small companies is to make decisions and be held accountable for these decisions. At large companies it is difficult to get some managers to take any risk at all. As a result there is little career growth because you never stretch yourself. At smaller companies, you get to make the decisions and you get to be accountable to management for your decisions. It is a great environment if you are a top performing professional.

Myth: Smaller companies don't act on opportunities the way a large firm can.

That's true, smaller companies often act much more quickly than larger companies. Instead of studying the problem to death and spending $20,000 in meetings to solve a $5,000 problem, they will assign the task to someone and let that person be accountable to make the correct decision. I remember when Ross Perot was on the Board of Directors at General Motors, he got so frustrated with the sluggishness of the board that he made the following comment, "If someone at EDS sees a rattler, he kills it. If a GM manager comes across one, he puts together a committee to discuss the situation."

Small companies tend to be more creative, nimble, and aggressive in pursuing opportunities. Fewer management barriers exist, and fewer voices are likely to chant "If it ain't broke, don't fix it," or "That's what we did twenty years ago." Small company management doesn't have the luxury of being king of the jungle. As a motivational plaque says, "When a gazelle wakes up in the morning, he has two choices: He can either run like hell or be eaten." Smaller companies have a lot in common with gazelles.

So if you are considering changing jobs, you may want to consider a smaller company rather than a larger company. You might be glad you did.

The Benefits of Getting Involved with a Professional Organization

Regardless of the size of the company you work for, there are limitations to growth opportunities. Volunteer-based organizations offer the growth opportunities that many workplaces simply don't have as viable options. This is why it is a good idea to get involved with your professional organization? Let's take the American Society of Safety Engineers (ASSE) as an example.

Let's examine the benefits. First and foremost, you get out of an organization what you put into it. If all you do is pay your dues every year, that's all you will get, membership in an organization of 32,000+ members. Let's say you even go to the monthly chapter meetings, maybe the Regional Professional Development Conference (PDC), or the national PDC. What do you get out of that? Well, let's add it up:

- **Learning opportunities**. You get to learn new things on topics you may not deal with on a daily basis. This is the primary thrust of ASSE. If you can come away a little smarter on a particular topic then ASSE has been successful.
- **Networking**. If your boss or coworkers ask you something about SH&E that you know very little about, you can call someone who does. Quite often, colleagues can give you

answers on the spot. Perhaps you found a contact at a chapter meeting who can also help you. If you can get back to whoever asks you the question with an informed answer, guess what, you will look like a true professional. SH&E is such a vast profession, no one can know about everything. So knowing people who are experts in other aspects of SH&E helps all of us survive. SH&E professionals are brokers of information and they give it freely. I am a member of numerous professional organizations, but only ASSE members give information so freely. It is one of your best resources.

- **Being informed**. This is different from networking, but is closely related. Being an involved member of ASSE allows you to hear things that are never put in print, things like, that standard will never get out of the Office of Management and Budget because of this or that, Or hearing that the Process SH&E Management standard is going to happen quickly so you better get moving. There is an old saying that there are three kinds of people in this world, those who make things happen, those who watch things happen, and those who wonder, what happened? If you choose the first category, you will definitely be more successful. If you are a consultant of some kind, being able to speak intelligently about current events in the SH&E business to potential clients is very important. If your job is SH&E and you don't know what is happening in your field, you don't look like a professional that really knows his or her profession.

Potential jobs. I've often heard someone say, "I think I'm about to be laid off so I need to concentrate on finding a new job." There isn't a more perfect place to network and find jobs in an expedient manner than ASSE. Many SH&E jobs never show up in the want ads because someone in the network found out about it and applied before the company had to advertise or go to a headhunter. Usually, these are by far the best jobs. Think about it from an applicant's point of view, if you apply for the job before the company goes public, there is little competition. If you apply for a job in the Sunday paper, you are competing with literally hundreds of applicants. You make the choice.

Furthermore, we've all seen fellow professionals show up at a professional group's monthly meeting, in some cases for the first time, saying, "I'm getting laid off so I thought I'd come to the meeting and see who is hiring." What a lame approach for one's career. Here they are, showing up for the first time to a chapter meeting with their hat in hand, looking for a job. If they do get a job, they usually disappear back into the woodwork. Would you hire someone with this attitude towards their career? I sure wouldn't. On the other hand, if they are involved, and the same thing happens, they are more likely to already KNOW who is hiring and more likely to find a job quickly and land on their feet.

- **Potential business**. I'm not talking about hitting your colleagues up for business at meetings; however, having a presence and just letting people know who you are and what you do can help. If you give a technical presentation related to what you do, you are educating rather than selling. Many SH&E professionals go to meetings looking for experts in particular areas. If you aren't there, how can you make yourself available? If I'm going to hire a consultant, I want to know a little about them and if they are members of ASSE.

Let's say you take the next step and volunteer for a committee or get elected to an office. What are the benefits of volunteering in this capacity? In addition to all of the items above, let's add up the new benefits:

- You are now part of the process. You will have an opportunity to see how things currently operate. If they aren't exactly to your liking, then you can be like the Boy Scouts and leave a campsite better than you found it. That is, now you can help improve the process.
- If you like what you are doing and are doing a good job at it, you may get to be a committee chairperson, or perhaps even get elected to an office. At these next levels you can learn how

your organization operates and network with more members. In my ASSE Chapter, I helped organize a symposium on SH&E software. When it was all done I knew four to five times more professionals, both software SH&E experts as well as professionals who wanted to learn about SH&E software. I also knew many things about putting on a symposium.

- Leadership. Believe it or not, the world is lacking true leaders, leaders with character and conviction. Just look around, leaders are hard to find. Your professional association is a great training ground for those who have that calling. An ASSE Chapter provides a great opportunity because there are so many people who can help you, such as past officers and other committee chairs. You can't go wrong.
- Enriching. Giving something back to an organization can be very enriching. To know that you have participated in making something better brings great satisfaction. It diversifies you and makes you a better professional.

So, what's stopping you? Take the next step and get involved. Volunteer! There are many benefits!

Communicating with Your Boss: Fire-Proofing Your Job

No matter where you are located in the food chain of your company's organization chart, getting *fired* is something that can happen to all of us. Regardless of the unemployment rate you can get fired due to downsizing, layoffs, mergers, acquisitions, and sell-offs. You can get fired as a result of your own actions, insubordination, fraud, misrepresentation, or a poor relationship with co-workers, subordinates, or superiors.

However, there are several things that you can do to avoid getting fired or being laid off, regardless of where you are placed in the corporate food chain.

The Early Bird Gets the Worm

Try to get to work before your boss gets to work, and if possible leave after your boss leaves. If your boss gets to work at 7:30 a.m., you get there by 7:15 a.m. If you work downtown and want to beat the traffic, get to work even earlier, perhaps 6:30 a.m. It is amazing the things you can get done before 8:00 a.m., when the phone rings and hallway conversations strike up. It also doesn't hurt to make the coffee for everyone, especially if your boss drinks coffee. He will most likely find out who makes that first pot. If he doesn't, make sure he does, however subtly. Getting to work before your boss gets there demonstrates that you are serious about your job and committed to it. Perhaps this is a good quiet time to meet with your boss and discuss many issues he would not discuss with you otherwise with an office full of people. Perhaps even a little bonding will develop, endearing you as much as a friend rather than just a colleague. For the mathematically- and loyalty-challenged, this means that you are probably going to be putting in more than eight hours a day. This goes a long way to fire-proofing your job.

Be Flexible

"Blessed are the flexible, for they shall not be broken" should be added to your personal list of beatitudes. Be ready to do something outside your job description, perhaps even outside your technological comfort zone. If you are asked to write a specialized SH&E report outside of your field of expertise, research it, find some models, call some colleagues, and figure it out. Don't say, "That's not my job!" or "I can't do that!" If you want to fire-proof your job and potentially move up the food chain, it is good to show that you are willing to learn a new discipline, especially if you are going to potentially manage that discipline.

Taking this a step further, periodically ask your boss if there is anything you can do to help him. Ask him for things you can do that he doesn't like to do or doesn't have time to do. That way you can learn what he is doing and be able to act more responsibly in his absence. Also, be vigilant when your boss is stressed and needs help. Helping him meet a critical deadline can pay tremendous dividends in your future raises, bonuses, and promotions. This isn't a guarantee, but your boss may very well speak well of you to others in your absence, thus furthering your career. If you can do these things you can easily add another number to the R-factor to your fire-proof rating.

Speed it Up

When you boss gives you a task to do, **STOP** what you are doing and get on it immediately. Your boss needs to know that the tasks he gives take priority over your own agenda. When he asks a question you can't immediately answer, research it and get back to him within the day or first thing the next morning. Your boss needs to know that you are committed to working for him and that you will sacrifice your own agenda to help him meet his deadlines. He is your customer first. Speed indicates that you are on top of your discipline and that it doesn't take too much time to find the answer he is asking for. At a minimum, check your work to make sure it is accurate and correct before submitting it to him.

Walk in Lockstep with Your Boss

This may not be popular, but agreement with your boss is very important. Your personal view may differ from your boss's view. However, it does not need to be aired in public, even if it is an ethical issue. Share your personal thoughts with your boss in private, but always support him in public. It is analogous to the principle of "praise in public, chastise in private" that many use with subordinates. Don't make your boss look the fool. It can only hurt you in the long run. Also, remember that he is still the boss and may want to run the organization his way. You work for him. Think about your future. If you want to be the boss some day you will be able to run things your way. Until then, be patient.

Put Your Boss in the Best Possible Light

Not all bosses are liked by all employees and co-workers. However, your boss most likely hired you or had a direct input into your hiring. You need to support and defend your boss when he is not present to defend himself. There are certain positive ramifications.

One, people will know that what they say about your boss will get back to him if they say it in front of you. What these people are doing is commonly called backbiting. Knowing that you are a direct conduit to your boss will force people to say less disparaging remarks about your boss in front of you. At one location where I worked, a co-worker would come into my office and tell me how bad the boss was. It was a touchy situation; this is the guy who just hired me! After a few sessions of non-committal responses on my part, I told my co-worker, "If you are looking for me to agree with you about how bad our boss is, you are barking up the wrong tree. This guy just hired me. If you have concerns regarding your job and your relationship with our boss, I suggest that you approach him rather than telling me your troubles. After all, telling me will not effect any change in him." Taking this approach will avoid any Human Resources problems in the future. If your co-worker's problems escalate, it may well become an HR issue. If you are later called in to the HR department to attest to past comments, you can say that you had no problem with your boss, rather than having to justify what you have said that may have corroborated with this employee's disgruntled comments. It is difficult to tell HR that you really didn't mean to say this or that, or you really didn't agree with your co-worker when he said this or that. Be careful—the job you save may be your own.

Two, it will build a sense of loyalty between you and your boss if he knows you will come to his defense in his absence, rather than feeding the fire. He will know that he can trust you to defend him rather than tear him down. Also, remember that these conversations sometimes take place in the hallways. The walls have ears, and what you say make return to haunt you. Negative comments travel faster than good. These comments have an uncanny way of making their way back to your boss. If they do it can become embarrassing and can definitely affect your future.

Also, if you step up and make the right decisions extemporaneously, that will make your boss look good. When you make your boss look good, it always pays off. Other ways to put your boss in the best possible light is to interact well with other people. You represent your boss, your department, and, when in public, you represent the company. It is a small world out there and disparaging remarks have a remarkable way of getting to people you never meant them to. If you can accomplish this, you can boost the R-factor several numbers.

Mind the Store in Your Boss's Absence

Before your boss leaves for an extended vacation, sit down with him and find out what the hot items are that may need attention in his absence. Make a list and check it during his absence to make sure you are poised for anything that happens. If those hot items surface, address them and write up a status report for your boss upon his return. Take the initiative to tackle and solve problems in his absence, while also getting your own job done. If you want to be promoted one day, this is will impress your boss's boss. Think on your feet. If something happens that appears to be beyond your ability check with other corporate resources and colleagues. Try to take care of business so that when your boss returns, his to-do list can be shortened.

Bloom Where You are Planted

Too many people today want their boss's job so badly that is all they talk and think about. They appear to be obsessed with being promoted. It can irritate your boss and his peers and perhaps even your peers. Enjoy what you are doing and don't focus on your next promotion. I have seen more people promoted because they focus on their job rather than trying to figure out how to find the "Up" button in their career elevator. I sometimes think upper management looks for those who are happy where they are when it comes to promotions and avoids those who want it all now. When someone asks me how I like my job, I reply, "I feel like a mole on Marilyn Monroe's face. I'm just happy to be here." Remember, getting promoted is a lot like traffic. Once you pass the car in front of you, you still have a whole bunch of cars ahead of you. Later on, that car you passed may end up passing you again anyway. So, don't worry about getting promoted. If you are a performer, others will notice and take the appropriate action.

Keep Your Cool

People who keep their cool have a propensity for making it to the top in the long run. They are unflappable and can handle emergencies as well as pressure from the media. The cool cucumber ultimately gets the kudos for making the company look like they have their act together in front on the public, other companies, and the media. Those who easily lose their temper often look the fools, making themselves look childish. Remember, it is not management by volume. Keep your cool even when you want to turn the volume up and your heart is racing. Keeping your cool can add value to your "R-factor" for fireproofing.

Zip Your Lip

If you are known as one who can't keep a secret, don't expect to make it to the top. To make it to the top, you must be known as someone who can be told anything in confidence. If you think you are impressing people when you share hot inside information, you are wrong. You will just look untrustworthy. If your boss discovers that you are the leak, he may decide to sanitize what he tells you, or not tell you at all, cutting you out of the loop. At one location where I worked, I discovered that I had a leak in the department. I had my suspicions which employee it was and set up a situation where I told only him a particular piece of information. I clearly indicated that my communication was in confidence. Once I heard that same information rattling around in the field, I cut him off. Later, he discovered that he had not been clued in on a major departmental activity that I wanted to keep under wraps until unveiled. He approached me and asked why he wasn't clued in. I told him that he had proven himself untrustworthy. It was quite an embarrassment for him. If this happens to you in the future, you may not be privy to some hot inside information that affects your job. Others will know more than you will. You just blew your opportunity to be on the inside. Don't be known as someone who shoots from the lip.

Look the Part

Wear the right clothing for the right occasion. When in the field, dress down in the appropriate casual attire. When in the office, match what your boss wears. Don't try to take fashion risks to make statements. Don't commit a fashion felony or be the GQ person in the office. Do be known for being a sharp dresser. It is kind of a conservative balance. I'm not saying you can't wear loud ties. Just don't wear them every day. Wear them occasionally on Friday (which is usually casual day). Remember the old saying, "Dress for the job you want, not the job you have," and you will be okay.

Be physically fit. Although this is not always true, those who are physically fit just present a more pleasing image. Have good grooming habits. This goes beyond combing your hair. Personal hygiene reflects on you, your boss, your department, and your profession. If you don't care what you look like, others may not think that you care about what you do.

Address Thorny Issues

In SH&E there are literally tons of thorny issues and potential land mines. Be current on the issues and technologies and be able to speak extemporaneously on them. Don't side-step the questions. Address them head on. Give no room for rumors to creep in. Think before you speak so that you do not misinform. Show that you know the issues and all related areas of potential impact to SH&E.

Anticipate

Never expect that everything will go as planned. Have a Plan A, a Plan B, a Plan C, and perhaps even a Plan D. This takes time and careful thought. Don't say anything to the troops that you wouldn't mind your boss hearing. Find out what your boss's questions would be and answer them before he asks them. Just like the Boy Scout motto, "Be Prepared."

Be Decisive

One thing SH&E professionals hate is decision-makers who waffle on making decisions. Don't be one of those people. Give the options thought and careful deliberation, but in the end, make a decision. Many people try to wait until the last possible moment to decide and then put it off because of new information. Make the decision now, but be prepared to alter your course based on additional information.

Keep Getting Smarter

Readers are leaders. To get to the top and stay at the top you had better keep learning every day. You need to be plugged in so you are clued in to what is going on in the SH&E profession. Being known as someone who is smart on the issues and constantly learning helps keep you fireproof.

Lighten Up

Everyone wants to be part of a positive activity. The converse is also true. Negative people tend to suck the energy out of you and give you a frontal lobe headache. Be the positive person that people like to see coming because you are a joy to be around, whether or not you are bearing bad news. Also, when it comes time to fire people, I can hear them now, “Aaaaaahhhh. Now I don’t have to listen to that naysayer anymore.” You’ve just been given a tenure track for the front door. Increase the R-factor by one.

Be Able to Speak on Your Feet

Most SH&E professionals speak well on their feet. However, this is not always the case. We all know how important it is to speak extemporaneously on a variety of SH&E-related topics. This will also help when you are called to speak to the Board of Directors or the company president. An example of speaking extemporaneously is the CEO approaches and asks, “In twenty words or less, tell me why we need an SH&E management system?”

Sing in the Rain

When you are handed a tough task, don’t drop your head and trudge off to the salt mine with a bad attitude. Lighten up your attitude and attack it like you would any other SH&E-related task. Don’t worry, be happy in all that you do. Bend over backwards to do it well. Don’t act as if it can’t be done or you don’t know where to start. “Just do it!” The way you act gives you an excellent opportunity to create a positive impression to your superiors. Who knows, perhaps your name may now come to mind when they think about the next vice president. After all, you’ve already shown that you can do technical tasks outside your stated expertise.

To survive the acquisitions, mergers, joint ventures, downsizing, rightsizing, bitesizing, and potentially capsizing, fireproofing your job is crucial. Keeping these things in mind can help you from getting fired in the long run by illustrating your corporate worth and keeping your boss happy.

Eight Ways To Get Noticed At Work

Getting the recognition you deserve can be a challenge in corporate America, especially at larger firms, where you're often competing for attention with hundreds of other employees. Here are some common-sense strategies for differentiating yourself from the herd.

Learn What Your Boss Values

Get to know your employer's style and play up to it so that you can present information in the format that's most likely to make an impact. If your boss is a numbers guy, for instance, quantify problems and possible solutions. Present your analysis to him in terms of the bottom line. Or, conversely, if you've got an employer who values vision, for example, he wants ideas on moving ahead in the marketplace, offer forward-looking suggestions on acquiring that competitive edge.

Study, Study, Study

Besides keeping up on industry news (a must for any employee), pay attention to trends beyond your particular specialty, so that when you contribute ideas, they're not limited to one narrow part of the market. This demonstrates to your employer that you're capable of thinking about the big picture.

Get to Know Your Boss

Learn a little bit about your supervisor so that when an opportunity arises to make small talk, you'll have something to say. Find out about her family, where she went to school, any hobbies she might have. If she's into a particular sport or pastime, for instance, read up on it. If you feel self-conscious, keep in mind that most people enjoy talking about themselves, so your boss will probably be pleased if you ask questions about her. (But don't overdo it. A little flattery never hurts, but you don't want your interest to seem contrived.)

Don't Whine

It's fine to bring problems to your boss's attention as long as you simultaneously present possible solutions. Otherwise, you risk coming across as a bellyacher. Your boss has enough difficulties, he doesn't need more. If you don't have a definite answer to the problem, at least suggest a starting-off point. Again, by presenting solutions to challenges, you show that you're thinking of the bigger picture rather than merely taking care of yourself. (Remember, everyone has problems at work, it's just that not everyone talks about them!) Ask questions. No one should be afraid to ask questions, and especially not new employees. Use questions as an opportunity to display your curiosity about your job and the company as a whole, as well as to demonstrate a bit of what you've picked up about the industry.

Be a Good, Solid Employee

This goes without saying. Be accountable for what you do, do your job to the best of your ability, and pay attention to details. Without these basics, the rest of your efforts to get ahead will likely go to naught.

Dare to be Different

A surefire way to get your supervisor's attention is by doing something a little bit differently from the rest of the pack. Start small by developing a personal style, add flair to reports you write and meetings you conduct by incorporating jokes or quotations you find memorable. This will make you stand out in people's minds.

Don't Be a Clock-watcher

No matter how hard you work, it doesn't look good if you make a mad dash for the door as soon as the whistle blows. Nowadays, many offices value workers willing to put in extra time. If yours is one, stay late occasionally, or take a shorter lunch hour. It may cut into your free time, but it will impress your employer.

Conclusions

A few simple ways to get ahead all deals with what you can do. Some of it is common sense, while some is just being a good employee, willing to earn and learn the job. Get this in your head and your heart and you will go far.

Success Tips

Organizationally

1. Strive to report to the highest officer in the company, preferably the president or CEO.
2. Develop a mission and vision statements supported by clear objectives.
3. Identify your role and scope in the organization.
4. Develop and execute a plan for continuing education.
5. Integrate with operations and build strong relationships.

Fireproofing Your Job

1. The early bird gets the worm
2. Be flexible
3. Speed it up
4. Walk in lockstep with your boss
5. Put your boss in the best possible light
6. Mind the store in your boss's absence
7. Bloom where you are planted
8. Keep your cool
9. Zip your lip
10. Look the part
11. Address thorny issues
12. Anticipate
13. Be decisive
14. Keep getting smarter
15. Lighten up
16. Be able to speak on your feet
17. Sing in the rain

Personally

Ten Commandments of Career Success

1. Know thyself
2. Know thine industry
3. Thou shalt network
4. Research thine options
5. Cultivate a good attitude
6. Think like an entrepreneur
7. Have a gimmick
8. Respect thine interpersonal skills
9. Set short- and long-term objectives
10. Feed thy soul

Professionally

Getting involved in a professional organization brings many bonuses

1. Learning opportunities
2. Networking
3. Being informed
4. Potential jobs
5. Potential business
6. Leadership opportunities
7. Enriching

Ways to Get Noticed at Work

1. Learn what your boss values
2. Study, study, study
3. Get to know your boss
4. Don't whine

5. Dare to be different
6. Don't be a clock-watcher

The Difficult Job of Hiring SH&E Professionals

One of the keys to career success is being able to find and hire good talent. Many SH&E professionals have a network to rely on to find talent. Often, that well runs dry. When it does, we are forced to seek and find SH&E professionals of whose performance we do not have firsthand knowledge So, when it comes to interview time, we must rely on the face-to-face meeting. Even before that, we need to know how to review resumes and narrow the field down. In the end, if we hire SH&E professionals who do not perform as expected, or even fail, it is a mark against our management ability to recognize good talent. The converse is also true: we can make a name for ourselves if we are able to recognize talent that provides a significant contribution to the organization and fits in well with the culture.

Interviewing

Whether you're hiring or just someone running a department, interviewing may be a key part of your job. It's not a task to be taken lightly. Although the job market is fiercely competitive, you don't want to move so fast that you hire the wrong people. You want make sure your interviewing practices don't lead to a lawsuit. Also, you don't want to waste time and money interviewing inappropriate people.

How the Pros Do It

Experienced hiring managers agree that interviews aren't just opportunities to question a candidate. They are also opportunities to listen, delve beyond the obvious, and discover if the person has the right skills and meshes with your company's culture. Many companies that hire permanent staff annually also spend at least three or four hours on the interviewing process per temporary or contract hire.

Most people who hire evaluate more than job-related knowledge. They also question how well the candidates understand the company's business and their abilities to contribute to its goals. Can they handle the stress and challenges of the position? Are they better suited to independent or to team structures? To a start-up or a mature organization?

Good interviewers know that interviewing can be costly, but it's even more expensive to hire the wrong people. They get the information they need from applicants, network to go beyond resume references, and don't waste time.

Your job is to present candidates with an accurate picture of what is going on at your company. Ask candidates if they've worked with other people or by themselves, if they can communicate and think on their feet. What's their attitude about customers? Do they regard them as a pain or someone who pays their salary? I also delve into the person's goals. If someone tells me they want to make $35 million by the time they're 35, I don't hire them. I look for people with goals the company can realistically satisfy. We tell them that, as soon as they get in the door, they're accountable.

Some people divide their interviews into multiple sections, asking applicants to talk about themselves and their work history. Then they ask very specific questions. Rather than asking what a candidate's compensation level is, they ask what their last W2 was. How would they be described by their current manager? What are their most significant accomplishments? What key areas would their manager say needs improvement? They ask them to describe the type of organization where they'd most likely succeed. At the end of the interview, if they don't have any questions, that sends up a red flag. I suspect they're not really interested in working here.

Planning the Process

Interviewing is a two-way street: you have to prepare, too. To capture the best candidates and avoid legal problems, you need a detailed, standard interview process and managers well-versed in what the company does, what it needs, and how to conduct a good interview.

Start first by writing a complete job description and forming a list of job- and experience-related questions to ask during the interview. Save time now: ask the appropriate managers to agree on the position's duties and role in the organization. Identify the job's key requirements in terms of education, work, and technical experience. Make a list of all essential behaviors the job requires (for example, must work in a team, must work in isolation, can work without direction, and so on). If tests are used, they must be administered for all applicants. Next, assign a point person to handle logistics, paperwork, and the scheduling of people and interview rooms.

One way to save time and money is to do in-depth applicant screening over the phone. But don't palm off this task to just anyone in the company. Make sure the phone interviewer has the skills to do this job properly. And when you finally invite an applicant in for an interview, tell him or her what formal tests or panel interviews are involved. Also mention the people he or she will be meeting and their roles in the organization. Be professional in your conduct and presentation. You want applicants to tell other people about the great interviews they had at your company and why they want to work there. Remember, every interview is like a sales call. The selection process is mutual. When the hiring process fails, it's often because interviewers haven't done their job.

During the interview itself, don't stray beyond questions emphasizing experience and business issues. Stay away from personal and non-job-related questions like, "What do you do in your spare time?" The million-dollar question is really, "How much work experience do you have doing X?" And when you wrap up the interview, say something like, "Thank you for your time. We have several applicants, and when we complete our interviewing process, we'll let you know if you've made it to the next cut."

Get Input

Now it's time for managers involved in the process to get together and rank the candidates, using a standardized format. If people are excited about certain applicants, ask them to define why. If no one is a good match for the job, ask yourself why. Are the job and salary expectations realistic? Do you need to rethink education or experience requirements? Are you losing candidates to the competition because of some unrecognized factor: your office environment, work hours, a difficult commute, or something as simple as not offering a candidate coffee or water during an interview? Do you have the right people involved in the interview process? Then look at the mechanics of your interview process. If you've spent too much time on people who aren't right for your company, learn how to improve your candidate qualification process and disengage gracefully when someone is obviously not the right fit. Finally, stay up to date on legal issues. Whether you're hiring executives or entry level employees, make sure everyone on your interview team asks the right—and legally appropriate—questions.

Resume Reading 101

How to Glean the Important Facts at a Glance

It's 15 minutes before your 10 a.m. candidate arrives for an interview. Your desk is strewn with paper. You flail about looking for that resume the boss gave you last week.

"Aha!" you cry with relief. "It's here!" The candidate is waiting in the lobby. You have six quick minutes to read the resume, remember why you invited this person to interview, revise the project's specifications, review your other candidates, write down the key questions you'd like to ask, realize who else will be interviewing her today, and recall who's taking her to lunch. "When's my next meeting?" you wonder, finally glancing at the resume. Enter candidate!

This scenario is all too common in offices across the land. In spite of the best-laid plans and a resume in hand for two weeks, most of us usually just glance at applicants' resumes without knowing what to look for and what to look out for.

Knowing how to decipher a resume (and a cover letter) is one of the basic art forms of the evaluation process. It's best done when you can devote your full attention to the task. You also need the right set of skills. You should be thoroughly familiar with the job being filled and know what skills it involves. You should know the buzzwords and which experience, training, and certifications are critical and which aren't. You should be aware of when a candidate's presentation is important, say, for a marketing applicant. Perhaps most important, you should be able to detect red flags or soft spots on a resume that need further exploring. Finally, you need a knack for deciphering who will be a good match for both the job and your corporate culture and why the match is mutually beneficial.

What to Look For

The physical. How people present their qualifications for a job on paper says reams about them. Is the resume easy to read or crammed? Look for a resume that's smartly designed and well organized. Look for a resume that states a clear objective. Failing that, try to find a candidate whose education and technical skills are well suited to your company.

A resume's look and organization is especially important for people applying for SH&E positions. It reflects their most current accomplishments, tools developed and implemented, programs implemented and results achieved. Do you see a lot of spelling and grammatical errors? They may be professional wizards or great on their feet, but what kind of presentations or documentation will they prepare?

Details, details. Savvy resume readers look beyond the buzzwords and dig for details: the experience, training, and track record that substantiates applicants' claims.

Look beyond what people know and ask how they have applied those skills. Some hiring managers say resumes should indicate clear results of developing, implementing, and maintaining a comprehensive SH&E program.

Consistency and quality count. Even in this age of shorter job stints and contract employment, look for resumes that show consistency, growth, and wise choices. Have the applicants worked for good companies on good projects? If they're contractors, did they last longer than two or three months? If they were permanent hires, did they work at each company for several years? Have they climbed the ladder of success? I like to see a progression of responsibility. Have the applicants gone from SH&E engineer to SH&E manager and stayed at one company for a while?

Motivation helps. Of course, not everyone will have cutting-edge experience. But resumes can hold clues to people's abilities to make the leap to your company. If candidates show energy, have done research, have gotten the training, and can show how their experiences in SH&E can help them bridge the gap, it can make a critical difference in decision-making.

Red flags. No resume is perfect. The art is to separate major problems from minor stumbling blocks. Are the job seekers local or are relocations involved? Are they single or do they have families? Do they need visas, and is the company willing to sponsor them? If the applicants lack degrees, will your hiring manager care, or is a CSP, PE, CIH, CHMM, REM, etc. enough?

Pulling It All Together

Good resume reading goes beyond looking for facts. It's also a matter of discerning the subtle factors that will make an excellent match—or spell disaster. Of course, the in-person interview is the other part of the evaluation process, but a resume can reveal much about candidates' desires, motivations, and abilities to fit into your corporate culture. Checking for experience is a no-brainer, but what applicants omit or how they shade the truth can often tell you far more. When you've mastered this art—and just as important, when you know your company's needs, culture, and peculiarities cold—you'll be on the road to making excellent matches.

Making a Good Hire Takes Instinct and Research

Armed with strategically-phrased interviewing questions, reference checking, and pre-interview resume reviewing, all coupled with your gut instinct, can get you the perfect hire. In the globally-competitive economy, it's more critical now than ever to ensure that you make a good hire that will last..

Before You Start

Before you start interviewing, define the most important skills, experience, character traits and knowledge you need. Make a list of attributes the applicant must have and stick to it.

For SH&E, typical skills include character, education, experience, professional society membership and certifications, communication skills, writing skills, and interpersonal skills. I look for a modicum of these skills melded with how the personality of the applicant will fit into my organization.

Education. Strongly consider a minimum requirement of a BS degree in a related field for an entry level position. Notice that a related field is acceptable. An art degree is not a related field. The degrees scale up as the level of the position requires. As an alternative, you may consider an applicant who has no degree but is a graduate of the school of hard knocks and a CSP. Sometimes these people are more productive and useful than those with degrees and no CSP.

Experience. Consider a minimum requirement of two years, either as a summer hire or a full-time employee in a related field. You may want to consider an education for experience swap. For example, you may consider swapping two years as a full-time student in a safety curriculum for two years of experience. However, you should exercise caution to ensure that the person you hire has a chance to succeed.

On the flip side is a candidate who has many years of experience and is applying for a position requiring only a few years of experience. At first blush, it appears that this person meets all the requirements. Exercise caution here. You will need to determine if the person will be happy in the position. If not, you and the employee will have a trying tenure. The bottom line, if they aren't going to be happy in the position, they won't be there long and it is likely that they won't be as productive, and may also damage the cohesiveness of the team.

Professional Society Memberships and Certifications. Anyone serious about their career should be a member of a professional society. Whether it is ASSE for SH&E, AIHA for an industrial hygienist, or AAOHN for an occupational health nurse, they should be a member in good standing. If they are further along in their career, they should also have or be able to attain a CSP, CIH, PE, or COHN certification. Contrary to what many believe, certifications illustrate a minimum, not a maximum, competency.

Communications Skills. I'm not saying that the applicant should have a long list of presentations, but he or she should be able to speak to a small group of people. You might want to schedule the applicant to speak to a group of employees at an SH&E meeting to get a feel how he or she can handle 'Joe Employee.' The applicant should be able to speak at the level of the audience and have a sense of genuine caring and truthfulness for the audience.

Writing Skills. Since we live in a world where everything is written down, applicants must have the requisite writing skills. If applicants write poorly they are marginally employable. If they cannot communicate with the written word, it is difficult to fit them in to any organization. Ask for samples of the applicant's written products, such as policies and procedures and technical publications.

Interpersonal Skills. One skill set that should be strongly considered is interpersonal skills. Once you determine that they meet the minimums of the criteria, look to see if they will fit into the personality and culture of your department and the company. This is a real intangible and needs to be assessed by asking hypothetical questions of how they would deal with people and conduct themselves.

Character. Character is a difficult thing to assess in an interview unless you know the delicate questions to ask. The things you want to assess are integrity, humility, loyalty, and ethics.

Integrity can be assessed in discussing your own views and then eliciting answers from the applicant. One example is while interviewing an applicant, we went to lunch and the applicant placed an item in his pocket and did not pay for it. It illustrated to me that if he was not faithful in the small things, how could he be faithful in the large things?

Paint a scenario where the applicant can either blame the employer or the employee as being the guilty party. If the answer continually comes up "the company," it may indicate the typical mindset today that "the company owes me." Remember, integrity is what you do when no one is watching.

If the applicant has a criminal record, ask about it and get him or her to explain in detail. Some surprising facts may surface. Another aspect is evaluating the appearance of impropriety. This means walking above the rules so that no accusations can be made as to whether you are in fact following the rules. It also speaks to moral character.

Integrity also speaks to the avoidance of conflict of interest questions. As an example, consider the following scenario: you are on a board that is selecting a contractor to perform a specified scope of work. Let's say that you and the board select a contractor, and after the fact, that the contractor approaches you to consult in an unrelated area. The best approach is to deny the opportunity and refer the business to a colleague for two reasons, one, it is a conflict of interest, and two, there is the appearance of impropriety. Even though the opportunity came up after the fact, the appearance of impropriety could ultimately come into question. Paint scenarios where issues like this are revealed and see where the applicant falls on the issues.

Humility. Humility can be explored by asking simple questions based on your own experience, things like, "When I was a youth, my parents required me to cut the grass and do chores around the house." If the applicant's reply is, "I didn't have to do that," continue by asking how he or she earned responsibility. If you come up empty, beware of arrogance.

Ask applicants how they might implement a new program. If they say that they got help and asked the people on the receiving end, it will be a good indicator of how they will perform. This is far better than writing a program and publishing it with little input.

Simply asking applicants the open-ended question, "Tell me about yourself," can be very revealing. Did they work their way up from the bottom and, as a result, have a respect for earning salaries, raises, and bonuses? Or are they considered an employee's right or entitlement?

Loyalty. Loyalty can be assessed by asking about situations with previous bosses. Did the applicant support and help his or her boss, or condemn the numbskull for boneheaded decisions? What are his or her aspirations? Does the applicant want your job tomorrow, even though he or she has little experience? Will the applicant be satisfied in this position or will he or she quickly outgrow the position and nip at your heels until you are done in?

Ethics. Ethics speaks to what extent the applicant will go to get his or her way. Paint a scenario that forces the applicant to think and evaluate right from wrong. For example, "You have an injury that some might argue does not need to be added to the OSHA 300 Log. What do you do?" Or, "You are conducting a due diligence of some properties your company wants to purchase. You notice a spill and discover that it has not been reported yet. Your company takes ownership tomorrow. What do you do?"

Interview Intelligently

Some techniques of intelligent interviewing are:

- Use tag team interviewers to get a well-rounded picture
- Let the applicant do most of the talking
- Elicit anecdotes about past behavior, not speculation on future actions

One of the best ways to find out how someone will manage pressure is to find out how he or she managed pressure in the past. The less adversarial you are, the more good information you get. You can also resolve the pressure issue with questions about past experiences and reference checks. Gather your data, but don't ignore your gut.

Intuition plays a role in hiring decisions, primarily in the nebulous realm of cultural fit. You have to ask the question of yourself, "Would I like to be around this applicant 40 to 60 hours a week?" But what determines if an applicant fits your culture is his or her past behavior, not just your instinct.

To determine whether applicants fit into a frugal spending culture, ask questions about how they handled such things as entertainment expenses in past jobs. I want to hear things like, "I treat expenses like I'm spending my own money." Once again, listen to your gut.

Evaluate Yourself as Well as the Applicant

Everybody knows managers who hired inferior prospects when better ones were available, or surrounded themselves with clones. Typically, these people make us feel more comfortable because they have a background similar to ours. Managers should measure applicants against the inventory of skills compiled, not against each other. If you don't, you might just get tired of the process and say, "Let's hire him, he's the best of the lot," which can easily lead to hiring a mediocre applicant.

Few job applicants can fulfill all of our wants, so decide which skills are critical and which can be overlooked or taught. Stop looking for pedigreed resumes. You can unearth diamonds in the rough at state colleges and universities and even smaller firms who just need an opportunity and some mentoring.

Don't Stop Recruiting

To ensure an adequate talent pool when you have an opening, stay active in your professional community. Look within your professional society for referrals. Ask your high achievers. Top performers tend to hang around with top performers; losers hang around with, well, you know, underachievers. This doesn't guarantee a perfect hire, but it can sure decrease the time between hiring replacements. It is difficult to assess the applicant in only a few hours, so look for information in key areas that can aid in the final decision to narrow the field. Interviewing and hiring people is a pain, and we don't want to do it unless we have to, so be careful that you make the good hire and don't have to repeat the tasks later.

Success Tips

Interviewing Basics

1. Look at the whole person not just job knowledge.
2. Clearly identify expectations if they are hired.
3. Ask probing, thought provoking questions.

Resume Reading

1. Look for a clean and professional resume without typos.
2. Look beyond the buzzwords for the meat of the accomplishments.
3. Look for consistency and quality.
4. Look for a motivated professional who shows energy and excitement.
5. Look for red flags to separate minor stumbling blocks from major problems.

Job Hunting, or "I Have a C-S-P, Now I Want a J-O-B"

Resume Writing Tips for the SH&E Professional

We just discussed hiring SH&E professionals. Now we will turn the tables and discuss getting hired. There are methods for evaluating resumes submitted to you; now we will take a closer look at resumes to help you in evaluating potential hires and improve your own.

SH&E professionals deal with volumes of data, and we are tempted to try and put all that data into a resume. The truth of the matter is that the resume is not a biography, but rather a marketing brochure that explains why to hire you and not the other person who submitted a resume. We spend too much time on position descriptions and not enough time on accomplishments. Read on to see how to a build a resume that gets the reader's attention and makes him or her want to interview you as the earliest possible opportunity.

Building a Good Resume

Building a good resume isn't easy. It's hard work and requires attention to detail. I actually keep two resumes, a long version and a short version. The long version chronicles all of my key accomplishments in detail. The short version is a combination of action statements that capture the long version accomplishments. I also maintain a list of publications and presentations. Every time I publish an article or make a professional presentation, I update my list. It is a wonderful tool during the interviewing process.

What I have assembled here is a list of tips for building a good resume that is an attention-getter rather than a trash-can magnet.

- For your first job, one page should do it, two is pushing it, but absolutely no more than three pages.
- Update it every time something significant happens.
- Include technical association memberships and significant activities.
- List professional credentials, such as PE, CPE, CSP, and CIH.
- Include the number of publications and presentations. If they are few, list them on the resume.
- Identify professional achievements, such as promotions and special tasks.
- List awards, both business and professional.
- A resume *does not* tell everything someone would want to know about you. It is not your autobiography or memoirs. It *does* tease readers to want to know more. It does have the minimum qualifications covered. It is an advertisement of you. It should be eye catching. It should make employers want to ask you in for an interview.
- Use verbs rather than nouns.
- Use demonstrated accomplishments or actions, rather than responsibilities or "I worked on X program."
- As a general rule of thumb, limit yourself to one page per degree. See guideline number 1.
- Avoid arcane titles.
- Describe the courses that you have taken that apply to the particular job opportunity.

- Always write a specialized cover letter to the individual point of contact. This is where most resumes fail. If you write a general cover letter, it will not necessarily work for every opportunity. You may even consider tailoring your resume for significantly different opportunities to highlight or explain particulars.
- Don't include unnecessary information such as hobbies or children's names and ages. Who cares? It's not relevant to job performance.
- Some jobs will ask for a salary history and desired salary. Include a salary history *only if* asked. *Always* put "negotiable" for desired salary. Make them tell you what they think you are worth first. They may think that you are worth more than you think you are worth. You may never know if you volunteer salary information. See "Interviewing" on page 63.
- You may need references. Contact your references and interview them before you give them to your potential employer. You want glowing, not so-so, recommendations.

The general elements of a resume include:

- Heading. Name, address, phone numbers, and email address
- Objective statement. What kind of position are you looking for? This can be tailored for various industries.
- Summary statement. One to two sentences summarizing yourself.
- Business experience. Self explanatory.
- Accomplishments statements. Self explanatory. You may combine this with the business experience section.
- Education. Your degrees. You may want to explain applicable courses here.
- Additional information. State that references available upon request.

Resume Pet Peeves

A recent survey by ResumeDoctor.com, an online resume consulting service, reveals recruiters' pet peeves about the resumes they receive.

ResumeDoctor.com surveyed more than 2,500 recruiters throughout the United States and Canada for the study. The recruiters were from varied specialties and industries, including engineering, information technology, sales and marketing, biotech, health care, administrative, executive, and finance.

"It really validated our thoughts that recruiters are receiving hundreds of resumes a day and that they see major problems with them," Mike Worthington of ResumeDoctor.com said. "It also provided valuable insight into the latest trends in Internet-based job seeking."

For example, applicants should avoid writing generic resumes that confuse recruiters.

"People often try to write a resume so generic that a reader has no idea what industry the candidate comes from," said Terry Cantrell of Panama City, Fla., an executive search recruiter.

"Did they manufacture fertilizer, package cow chips, cook and distribute potato chips or assemble computer chips? ... I am usually looking for a reason to exclude resumes, not a reason to include them."

As part of its findings, ResumeDoctor.com compiled a list of 20 things recruiters hate about resumes. Here is a sampling of the actual responses from recruiters.

Top 20 Resume Pet Peeves

1. Spelling errors, typos, and poor grammar.
2. Too duty-oriented; reads like a job description and fails to explain what the job seeker's accomplishments are.
3. Missing or inaccurate dates.
4. Contact information is omitted or inaccurate, and the email address is unprofessional.

5. Poor formatting—bad use of boxes, templates, tables, headers and footers.
6. The resume is functional, as opposed to the preferred chronological form.
7. The resume is too long.
8. Paragraphs are too long where bullet points could have been used.
9. Candidate is unqualified.
10. Includes personal information not relevant to the job.
11. Employer information is not included and/or does not tell in what industry or with what type of product the candidate worked.
12. Lying and misleading (especially in terms of education, dates, and inflated titles).
13. Has trite introduction or meaningless objectives..
14. Poor font choice or style.
15. The resume is sent in PDF or ZIP format, is faxed, or on the web, rather than emailed as a Microsoft Word attachment.
16. Contains pictures, graphics, or URL links that no recruiter will call up.
17. No easy-to-follow summary.
18. The resume is written in the first or third person.
19. Gaps in employment.
20. Important information is buried.

Additional Tips Regarding Resumes

Never volunteer your resume. You should be a human resume, with all of your significant accomplishments on the tip of your tongue, poised for the probing question, "So, tell me about yourself."

Never refer to your resume. When speaking, never say that the answer is in the resume. The interviewer obviously did not read it. You are just pointing out the interviewer's inadequacies and appearing to take control of the interview. Bad move.

Never change your resume during the interviewing process. So many of us think if we add this or tweak that, then our resume will get better, especially if a colleague or interviewer suggested it. However, the end result is a voluminous resume. The goal of a resume is to get someone's attention, not divulge your biography. The interview is the selling opportunity.

Smooth Sailing on the Seven Cs

A help-wanted ad in today's competitive job market may attract hundreds of responses. Your cover letter and resume must attract the reader's attention quickly, and stand out from among the pile of other competing resumes in a positive and discriminating manner.

You can almost guarantee this will occur by using certain new, yet proven, techniques. Simply help your prospective employer discover how he can save time and money by hiring you instead of any other candidate. It's easy to do and the results are positively terrific.

The likelihood of your correspondence accomplishing this will be improved by following certain guidelines. By implementing these "Seven Cs," your written communication will be more effective.

Creative

Your letter arrives with all the other mail your prospective employer receives that day. He has his "business hat" on, with many other issues competing with your message for his attention. Consequently, the message you *intend* the person to receive is not always the message he actually receives. Use inventive techniques to make yourself stand out and present your capabilities and benefits in the most positive, yet appropriately innovative manner.

For instance, you may come across an ad that requires you to send your compensation history with your resume. It's best not to divulge your salary requirements until after you've had the chance to meet with the potential employer personally. A creative way to address this issue would be to present a graph of your salary history as a function of percentage increases over time. Perhaps even mapping salary against cost savings and incident rate reduction will illustrate your positive impact on the bottom line. In this way you're not ignoring his request for salary information, but you are providing it in a way that is more conducive to reaching your objectives.

Credible

In order for your candidacy to be taken seriously, your creativity cannot be so outlandish as to reduce your credibility. Something that is true is not always believable. It may be true that your actions added value to your previous employer's stock and bottom line, or reduced costs by 50 percent. However, the reader may view this skeptically, unless you can provide proof to document this statement.

If your claims border on the outlandish, be prepared to back them up with facts, numbers, and references upon request. For example, a letter from a previous supervisor attesting to your skills and performance is worth more than its weight in gold. Place it in your portfolio and produce it upon request.

Convincing

In order for your cover letter to be compelling, it must document the truth persuasively. It should substantiate the reasons why you would make a better employee than any other candidate.

Use active words. You make your meaning and sentences more clear when you use active, action-oriented words. Instead of "the excellent results I have achieved have been the result of..." say, "I was..." instrumental in decreasing the incident rate by 50 percent through..." Begin your sentences telling how you, "managed, created, directed, and/or motivated," things or people to achieve a specific result.

Tailor your message as closely as possible to the reader's needs. Let him know you've taken the time to consider applying to his company carefully. Address what you understand to be his particular situation, and why your talents qualify you as the person who can help solve his problems.

Personalize your letters with the recipient's name and use it occasionally throughout the letter. If your correspondence is addressed to "Dear Sir/Madam," it will be relegated to the junk mail category and not given serious attention. It is more likely to be read and acted upon if you quickly get the reader's attention and give him a clear reason to continue reading.

Use positive words. Certain words have proven to be effective in eliciting a positive response. If you use the following words to convey your message, you'll be more likely to stimulate favorable action: *you, free, discover, SH&E, help, results, money, save, guarantee, health, new, proven, love,* and *easy.* Reread the first paragraph of this section to see how many of these words were used in it.

Promise a reward to the reader. Let the reader know in the first paragraph that if he continues reading, he will be rewarded. Brag-and-boast sentences usually turn a reader off before he gets into your message. Your promise should be specific: "Here's what I can do for you." Provide evidence that your claim is valid.

Talk person-to-person. Speak to the reader as an individual, for example, like one friend telling another friend about a good thing. Your writing style should be simple: short words, short sentences, short paragraphs, active rather than passive voice, and no clichés. Use the pronoun *you.*

Get the reader's attention. In advertising and newspapers, it has been shown that five times as many people read the headline than read the whole article. Also, each succeeding paragraph has progressively fewer readers. Therefore, make your first sentence function as a headline and grab the reader's attention. Then write each succeeding paragraph to keep him interested in reading further.

Begin your letter with an attention-getting statement. Demonstrate that you are familiar with the requirements for the position, and that your qualifications are well-suited for it. The first paragraph should be a one- or two-sentence description of the value you bring to the company and why it would be in the company's best interest to interview you.

Emphasize what you can do, not who you are. At this point, communicate only what will make the reader interested in interviewing you to learn more about what you can do for the company.

Complete

The objective of your correspondence is to obtain an interview. If you tell too much, or not enough, you reduce your chances of that happening. Present your value and show how your accomplishments meet your prospective employer's needs. Your goal is to hook the reader to ask for more detail. If you tell the whole story, then the reader won't need to hear the rest of the story.

Your cover letter must be complete, without errors of omission or commission. You shouldn't tell your entire life story in your cover letter. If you insist on telling a detailed account of all your background and interests, you may annoy the interviewer with extraneous facts that he or she considers a waste of time. Instead, provide only the information needed to raise your prospective employer's interest without being misleading. Do not intentionally omit critical information, either. Tell what is necessary and leave the explanations for the interview.

Current

In addition, your cover letter must be up to date. You can do this by keeping a diary of your recent accomplishments. It may be an award for incident rate reduction, or that you were a guest speaker at the local American Society of Safety Engineers meeting. Make the connection between what you have done in the past and what you can do for the company *now* by placing emphasis on current events.

Clear

Your cover letter will be ineffective in gaining the reader's attention if your message is not clear. Don't bury your important words in clichés and rhetoric. Briefly state what you want to occur and why it's to the reader's benefit to hear what you have to say.

Concise

Be succinct, presenting your skills so the reader immediately sees the relation between the company's needs and your ability to meet them. Just before Abraham Lincoln gave his famous Gettysburg Address, Edward Everett gave a two-hour oration. Lincoln's was written on one page and took less than ten minutes to deliver. Yet there is hardly an educated person in the United States today who cannot recite at least the first line of Lincoln's speech.

Your message should be as concise and to the point as the Gettysburg Address. Don't waste time warming up with extraneous information

Your prospective employer is not interested in what he can do for you, but what you can do for him. Demonstrate how you can solve his problems and make the company more successful as a result. Do this as briefly and creatively as possible. Follow the Seven Cs of effective communication and you'll be successful more quickly.

Finding the Right Look for a Job Interview

It doesn't matter what you wear on a job interview—whether you come in with old sneakers and sweatpants or a well-tailored suit—you will be judged solely on the basis of your impressive skills and sparkling personality.

Wrong.

There are times to dress for comfort, times to express our individuality via your Dilbert T-shirt, nose ring, and orange hair. Your job interview is not one of them. When your potential boss is looking over your resume, do you want her to be thinking, "Did I see gum on his shoes?"

You only have one chance to make a good first impression. The first impression you make will have a lasting impact, so effort spent on grooming and dressing appropriately has a definite payoff. Here are some tips for dressing for interview success.

Dress in harmony with the way those interviewing you are likely to dress. In general, interviewers are more likely to perceive you as being one of them to the extent that you are dressed in the manner of company employees. If you're uncertain about your potential employer's dress style, visit the company location as employees are arriving for work. A good rule of thumb is to wear clothing just a little dressier than everyday work attire.

Don't wear something so striking that attention is drawn to your clothing rather than you. This almost always means conservative, traditional, and conventional clothing. Clothing that is faddish or has exceptionally bold patterns or colors attracts attention—unfortunately, away from you. Save those loud ties or sweaters for later.

Pay careful attention to details. Small, simple annoyances like unshined shoes, messy hair, jangling gaudy jewelry, mismatched clothing, overly prominent accessories, or untrimmed nails, often result in strong negative reactions by interviewers.

Purchase high-quality clothing that you know looks good on you. If you buy your outfit from the discount store you might end up looking like a sack full of doorknobs. This may mean buying an expensive new outfit, but it's a worthwhile investment. When you are buying interview clothes, consider getting all the appropriate accessories at the same time. This usually is the most effective way to produce a coordinated, impressive appearance. If you are uncertain about how to put it all together, make your purchases at a good shop with experienced salespeople. Stores that sell top-quality clothing usually have someone on hand who is good at coordinating a conservative business ensemble.

If you have to travel by plane to the interview, the ideal way is to fly in the night before. This makes it possible if necessary to have clothing freshly pressed at the hotel.

Do not wear anything that connects you with a specific political cause or affiliation. This precaution is simply a matter of minimizing risks. There are more biases out there than there are interviewers who share your beliefs and preferences.

It is often better not to carry a briefcase into the interview. It's just one more thing to distract you and to remember to pick up upon leaving. However, if your resume or other papers will be needed, select a small, high-quality case. Leather envelopes are ideal. With larger briefcases, getting at and finding material can often be awkward. Thin envelopes make it easy for you to reach in and extract whatever material is needed.

In the final analysis, your total approach should be one of neatness. Many interviewers interpret a candidate's disheveled appearance as a sign of disorganization and carelessness.

Interviewing Tips for the SH&E Professional

The first step in preparing for a job search is to prepare an interview-specific resume. One version of your resume just cannot serve as the universal document in a job search situation. Prepare a resume for each category of work situation. To create these resumes, research the individual company's business record, history, and the position offered. After learning the requirements for a certain position, design a resume that acknowledges each required skill and area of experience advertised. These resumes should be based on different industries and what they require, as well as different positions. For example, resumes developed that specifically highlight industrial SH&E, process SH&E management, risk management planning, and/or ISO 9000 would be used for companies with these core attributes. Also, resumes should reflect different levels of responsibility, such as corporate versus plant experience, multi-site versus single-site responsibilities, and national versus multi-national locations. Carry extra copies of these customized resumes to the interview, even if the human resources department has a copy.

Researching the company before going to the interview is a valuable technique. Find out as much as possible about the company before the interview. Then, capitalize on this research during the interview by demonstrating knowledge of the corporate goals, recent major accomplishments, and stock performance. But don't overdo it.

Good resources for this research are the Dun & Bradstreet or Standard & Poor's directories. Another publication available through most libraries is *The Value Line Investment Survey*, a summary and analysis of the operations and financial accomplishments of companies. If no information about the company is available, don't be afraid to call the human resources department and ask. A candidate aware of the company's bond rating or recent changes in the price of stock may impress an interviewer. Additional information about a specific company and its operating standards may be found in the company's annual report or other public documents describing business activities.

Knowing about the company culture—how that company approaches particular situations and goals—is also good information to have prior to the interview. Use whatever resources are available: colleagues, bulletin boards while you are waiting for the interviewer, the annual report, the Internet, and other public domain publications. Know the rules of the company and review their procedural guidelines. Follow any established corporate protocol, particularly that suggested by the interviewer. By following rules to the letter, the first impression you leave shows attention to detail. Compliance with a company's culture also is important when selecting an interview wardrobe. Once again, follow the rules. There have been real-life instances where a candidate's appearance figured prominently in the hiring decision.

For an office or corporate position, wear conservatively business attire. For a site job, conservative business casual is suggested. A good guideline for any job interview candidate is to wear attire appropriate for the first day on the job. If you aren't sure, ask the interviewer what the proper attire would be for the interview.

When planning, set aside ample time for the interview. Give the interviewer plenty of time to work through his agenda. Appearing edgy or in a hurry can create a negative tone. If the interviewer feels rushed, it can be costly. A quick glance at a watch or any other sign of anxiety may give the impression that this candidate isn't focused on the interview.

Be punctual. A good policy is to arrive at the interview 15 minutes prior to the appointment. Walking into the interviewer's office late establishes a negative first impression.

A job candidate also is advised to maintain a positive frame of mind. In the interview, avoid topics about personal inconvenience or problems. An interview that begins on a down beat will create a poor impression that will be difficult to change.

It is also important to arrive at the interview alone. The presence of a third party, such as a friend, spouse, or family member can distract the candidate and the interviewer. This type of distraction may make a difference in the interviewer's assessment of the candidate's abilities and change the tone of the interview. Remember that you are powerless until the offer is made. You are a guest in their house, so act like one. If you were a guest in the house of someone you just met, you would never think of going to their refrigerator and pouring a glass of milk, would you? The same is true in an interview.

What are the Biggest Mistakes People Make in Interviews?

The biggest mistake professionals make in an interview is seeing it as the end of the process instead of the beginning of the process. You should figure out what you need to do after the interview to prove to the hiring team that you are the right person for the job. You have to be good at sales, which includes selling yourself, selling the product (your skills and capabilities), and selling the cost (your salary requirements). It also includes convincing the interviewer that:

- You are a professional.
- You have no problems.
- You will create no problems.
- You will solve their problems.

You've got to fight for the job. For example, let's say you interview at a place where they really need somebody to manage their Process Safety Management (PSM) program. Instead of saying, "Trust me, I can do it," what you can do is provide them with some of the programs you've developed and audit results (sanitized to be generic and not company-specific). Even better, bring a portfolio to the interview hidden in your briefcase, ready to produce if necessary.

One of the biggest mistakes is just not being yourself. I've often found that people who try creating a facade of what they would like people to believe tend to come across as phony. And as a result, it starts causing the interviewer to start marking negatives rather than positives about that person.

But you have to walk a fine line between being yourself and being what the company is looking for. A person who's very open and honest, even to the point of saying, "I've never done anything like that before but I'm a quick study," and selling the qualities of who they are rather than the experience they have, goes a lot further than somebody who tries faking that they can do the job or trying to answer questions based upon what they think the interviewer is looking for.

I think one of the things that comes across very strongly in any interview is how much a person is excited about the type of work the company is offering. And that projection has a very strong impact in hiring in the first place.

So I recommend that professionals who are looking for a job do some research. This research normally is to contact a company that offers certain products or services and, in so doing, has certain kinds of job opportunities. Also ,talk to the people who do that kind of work.

The least effective way is to find a job opportunity, go in and interview, and hope you get the job without any research, without any thought process, and without any related skills, talents, creativity, past experience, and knowledge. (And by past experience, I also mean having a family member who was in that field, having held a part-time job during school that had some exposure to that field, and those kinds of things.)

Just going out and interviewing for a job because there happens to be a job opportunity, not being sure what the job entails, and whether you would even enjoy the job once you're on it is not recommended.

The worst thing you can do in an interview is to say, "I'd be interested in X or Y or Z," instead of saying, "The major contribution I could make to this organization would be as an X." You have to be very, very focused. When the interviewer says, "What do you want to do?" that has to be answered in fewer than a dozen words.

Other common mistakes include:

- Failure to look the interviewer in the eye
- A limp handshake
- Arriving late
- Asking too few questions
- Intimidating the interviewer
- Poor oral skills
- Lack of career planning
- Making excuses, evasive answers
- Lack of maturity
- Cynical attitude
- Laziness

Interviewing Etiquette

You sit down and the interviewer comes in and offers you coffee. Do you take it or not? And the answer is, take the coffee, whether you drink it or not. Huh? You just said to be honest. Well, the reason the interviewer offered you a cup of coffee is probably because he wanted a cup of coffee. Now he can't have one. You just took control of the schedule. Remember, let the interviewer always have control. If you don't drink coffee, just let it sit there and get cold. No big deal, and the interviewer is in control.

What if the interviewer asks if you want a look around? Do you do it or not? Just as before, take a look around. Allow the interviewer to remain in control. This goes down to the smallest detail, like letting them push the buttons on the elevator, not reminding them that they passed the fourteenth floor, where the other person who was going to join you for lunch is located. Even though you remembered, let the interviewer remain in control.

Okay, what about shaking hands? Do you offer yours or let him offer first? Again, let the interviewer initiate the handshake. Something as simple as a handshake could set you off on the wrong foot. Even worse, what if the interviewer mispronounces your name? Do you correct him? No. Let the interviewer figure it out. Any attempt to take control of the interview is bad news.

If the interviewer goes to the rest room, you go to the rest room. Follow his lead. In short, mimic the interviewer's activities where possible. If you don't go to the rest room, you leave an awkward situation with you standing unaccompanied, waiting for the interviewer

Meal Etiquette

All right, you are going to get a nice meal after traveling across the country. What a treat! Hold'er, Newt! First, remember that the only one who may get a chance to enjoy the meal is the interviewer, not you! The meal is to determine your social graces at a minimum; at a maximum, it is to grill you over a meal. I remember when I was starving and we went to lunch. I got a huge lunch, but was unable to finish any of it. Remember, you will most likely not get to eat at all. Just keep that in mind during the meal.

Okay, you just sat down and the interviewer says that the steak and salmon are great. What do you do? Don't do the salad. Select the steak or salmon. Still, remember that you won't get to eat it anyway, and you are following their lead. What if the interviewer has no preference? Ask him what his preference is and then order one of those. Simply put, follow the interviewer's lead. This is neither the time nor the place to be innovative.

How to Make the Interviewer Like You

Many professionals approach the job interview with the idea that the main thing they must do is impress the interviewer with their accomplishments and qualifications. While these factors are certainly important, you must be liked by the interviewer. Obviously, the employer wants someone who is qualified for the job. Citing your accomplishments tells the prospective employer what you have done for previous employers.

However, the single most important element of the job interview is making the prospective employer like you. If the employer does not like you, it won't make any difference how good your qualifications are. Someone else will be hired.

How do you get the employer to like you? The best way is by listening to clues as to what he or she wants, and then trying to be that person insofar as you can within the boundaries of truth. Consciously or not, most employers tend to hire in their own self-image. They are looking for someone with whom they feel comfortable, an individual they believe will fit well into the company. If you can establish a commonality of interest with the interviewer, it is a plus.

If you cannot find this commonality, at least be careful not to give the wrong answers. For example, if the interviewer asks, "Do you think the Astros will win today?" your answer obviously is not to say you do not like baseball. The interviewer has given you a clue that one thing he or she wants is someone to talk baseball with at lunch. At the same time, do not pretend to be an Astros fan when you are not.

Be careful not to ask too many questions. Interviewers do not like to be put on the spot. And besides, all of your questions will be answered soon enough if the company decides it likes you and makes an offer. Remember, your objective is to get the job offer. You do that by making the prospective employer like you.

How You Answer Some Hard Questions During Your Interview

They key to success or failure in a competitive employment interview often hinges on how you answer some hard questions.

Whatever the questions, there are no easy ones in a job interview because it is the exact opposite of a casual conversation. The most difficult for most people being interviewed are the open-ended questions. These are filled with danger for the job seeker if you are not prepared with the appropriate responses. The following are some hard questions and some suggested responses.

Changing Companies

Question. "Why do you want to leave your present company?"

Answer. "I've been waiting for an opportunity like this for a long time. When I saw it, I just couldn't pass it up."

Advice. This query is usually the most difficult and emotional question of all for the person who has been discharged. It can cause embarrassment if you are not prepared with an answer and be unnerving to the extent that it ruins the remainder of the interview for you because you realize you are not making a good impression on the interviewer. You have to find an answer that is true, but one which does not reflect negatively on you, the company, or anyone in it. You want to avoid giving the impression that anyone was in the wrong. The explanation should be brief, no more than two or three sentences. Then stop talking. Do not attempt to elaborate or you will only raise additional questions that you do not want to get into.

Interest in This Company

Question. "Why do you want to come to work here?"

Answer. "This company is in a perfect position for me to do what I do best. My forte is bringing just the right level of structure while still allowing freedom to do SH&E and seamlessly integrating into operations."

Advice. Most of the jobs that professionals accept have never been publicized. Many have not yet been created when the professionals comes to call. If you try to answer specifically how you can fit in, you may be placing limitations on yourself. Instead, tell the interviewer how good you are at what you do and demonstrate that you are so well qualified that the company cannot do without you. Let them figure how you can best fit into their plans.

The strategy for success in an interview is to be whoever the interviewer wants you to be insofar as possible within the scope of your talents. You achieve this objective by listening for clues as to what he or she wants to hear and responding with the appropriate answers.

What all interviewers are seeking is the answer to the basic question of why they should hire you, as opposed to around six other candidates who are equally qualified. How well you address yourself to their image of the ideal candidate will usually determine whether or not you get the job offer. Obviously, you must stay within the boundaries of truth and not make the mistake of misrepresenting yourself. Employers are screening more intensively now in the wake of widespread concern about ethics in the workplace, so any personal misrepresentation on your part is bound to backfire sooner rather than later.

Your Strengths

Question. "What is it that you do best?"

Answer. "In my years of experience, I am best at relationship-building and problem-solving. For example..."

Advice. Answering this question can make you or break you as far as getting a job offer is concerned. What the interviewer really wants to know is what you can do for the company, and why you should be hired over several other professionals who are equally qualified.

In connection with this, the interviewer wants to know not only how good you are, but also if there is anything about you that could cause problems. Avoid the latter and cite specific examples of things you did in your last job, taking full credit where credit is due for programs initiated and work accomplished. In your advance preparation for this vitally important question, you should list the points you want to make and commit them to memory. Also, by concentrating on the specifics of what you have done, you are using the technique of bragging successfully without coming across like a braggart. Many professionals interviewing for jobs fail to win the offer simply because they fail to sell their accomplishments. Except for entry-level candidates, the employer will be far more interested in your work record than your educational background. You should mention your education, but discuss your work-related accomplishments first.

Your Weaknesses

This is the flip side of the previous question. When interviewing with a company, you will often be asked not only about your strengths, but also about your weaknesses.

Question. "Looking at your own resume, what do you think your weaknesses are regarding this job?"

Answer. "I believe that my skills and abilities are a good fit for this position. Do you have any specific concerns?"

Advice. Turn the question around and get the interviewer to disclose what he or she believes are your weaknesses. Then use the opportunity to change the interviewer's mind. Give specific proof. This is one of the old, stereotyped questions in job interviewing. For the professional who is well prepared, it is a welcome signal because it indicates the interviewer is a non-expert at interviewing. This gives you a chance to turn the interview to your own advantage. In answering the question, you should concentrate on the strengths and avoid the weaknesses. The important thing to remember is never to say anything negative about yourself. There is often a tendency to do this because we are taught from childhood that we should be modest in our dealings with others. But that sort of attitude has no place in a competitive employment interview. Even a seemingly harmless statement such as "lack of patience with inefficiency" is dangerous. It can be read as a sign that you have a quick temper, are hard on subordinates or cannot handle a difficult situation without losing your cool.

Your Ideal Job

Question. "What is your ideal job?"

Answer. "I really enjoy (use something from the job description that you know you can do well, such as leading Hazops) and seeing the improvement of the processes while ensuring safety and environment parameters are met."

Advice. What you should do here is to describe the job you are applying for insofar as you know. Whether or not it actually is the ideal job depends on a number of variables, but your objective is to get the job offer. After that you can decide whether you want to accept it, following negotiations on salary, benefits and other matters. If you talk about a job that obviously is not the one you are interviewing for, the interviewer will conclude that you are not interested in the available job and will drop you from consideration immediately. Be careful not to talk yourself out of the job in this manner.

Going the Extra Mile

When interviewing with companies, you will often be asked questions that seem straightforward to answer. However, an answer of yes or no should never be adequate. Always back up any statements you make with specific examples. This validates what you are trying to convey.

Question. "Would your current boss describe you as the type of person who goes that extra mile?"

Answer. "Absolutely. In fact, on my annual evaluations, she writes that I am the most dependable and flexible person on her staff. I think this is mostly because of my ability to juggle and prioritize."

Advice. Share an example that demonstrates your dependability or willingness to tackle a tough project. (If you describe "long hours of work," make sure that you prove the hours were productive, and not the result of poor time management.

Stand Out From the Crowd

Often in an interview, you will be asked to separate yourself from other candidates who may be more qualified or may be less of a risk factor.

Question. "What new skills or ideas do you bring to the job that our internal candidates don't offer?"

Answer. "Because I've worked with the oldest player in this industry, I can help you avoid some of the mistakes we made in our established markets."

Advice. This question addresses your motivation in adding true value to the job. Evaluate the job carefully, considering current limitations or weaknesses in the department and your unique abilities. Your ability here to prove "I offer what you need and then some" could land you the job.

Be Specific

Sometimes in interviews, you will be asked questions that lend themselves to be answered vaguely or with lengthy explanations. Take this opportunity to direct your answer in a way that connects you with the position and company, and be succinct and support your answer with appropriate specific examples.

Question. "Why did you choose this particular career path?"

Answer. "I chose aerospace because I was interested in the space industry throughout my childhood. After seeing the catastrophic Challenger and Columbia events my desire was to be part of the team to prevent those types of accidents from occurring."

Advice. You need to convince the interviewer that their industry and your career goals are in sync. Do you have a realistic view of what it is like to work in their industry? What aspects of their industry are particularly attractive to you? Give specific examples that the interviewer can relate to and convince he or she that this career path makes perfect sense for you.

Previous Employers

Question. "Tell me about your current and previous employers."

Answer. "I really learned a lot from implementing the Crisis Management and Emergency Response program. After that we implemented a work permit and contractor safety program that ended up saving the company about $2 million in potential losses."

Advice. Beware! Do not criticize current or former employers because it will reflect unfavorably on you. But you should not go to the other extreme and give your supervisors all the credit for your professional development. You should take as much credit for what you have done as you can, because this is what impresses the interviewer. Emphasize particular examples of initiative and leadership, where you created or led a project or program to successful conclusion or devised some new contribution toward improving the company's profitability. The best way to impress a prospective employer is to talk about how you improved profitability for former employers. So, if you were able to show your SH&E savvy as it related to dollars and cents (or sense) it will prove useful here.

Your Opinion of Your Present (or Previous) Boss

This is another trick question. This query is another variation of the last question. The same rule applies: never criticize a former employer, regardless of what you really think. That being said, oftentimes in an interview, questions will be asked trying to gain some insight about your past working relationships with co-workers or employers

Question. "Tell me about your relationship with your previous bosses."

Answer. "My bosses would tell you that I've often been a sounding board for them. With all of my bosses, I developed a close rapport and trust."

Advice. The interviewer is looking for a fit between the two of you. As you describe each previous boss, the interviewer will be making mental comparisons between your old bosses and themselves. You must be honest without being overly negative. Emphasize the type of boss you work well with.

Tough Questions About Your Past

You may have something from your past or on your resume that could present a sticky situation in an interview. Answer inquiries about your situation cautiously and try to come up with a creative way to turn a potentially negative experience into a positive response.

Question. "Why didn't you finish your studies?"

Answer. "I decided to leave school because I was working 30 hours a week waiting tables to support myself. I felt that I did not have enough time to devote to my schoolwork. When I do anything, I always give 150 percent."

Advice. The interviewer is trying to find out if there is a major issue that could interfere with your work. Do you tend to complete things?

Did you flunk out? Give a good reason why you did not finish or explain why any issues related to it are in the past.

Great Expectations

Let's say that you already have a job, but are looking for something that is more satisfying to you intellectually and in regards to your acquired skills. Employers want employees to be satisfied at the end of the day, to feel like they have accomplished something when the 5 o'clock whistle blows. Now is the time in the interview to let them know what would make your career more challenging, and what direction you want your career to go. Let's face it, if a company can't offer that challenge, you should continue your search for something that will.

Question. "What do you really want out of your next job?"

Answer. "I'm really interested in working for a company with vision, where I can use my background, experience, and skills to participate in the growth and expansion."

Advice. Interpret one or two items from your current work experience that explain why you are talking to the interviewer. Focus on limitations in growth or learning from your current job. Make sure you point out why you feel the job at hand provides the additional responsibilities you are seeking.

What You Can Contribute

Before an employer makes the decision to hire you, he or she will not only want to know your past performance history, but what other special contributions can be expected from you in the future.

Question. "Tell me about a special contribution you have made to your employer."

Answer. "In my last job I worked diligently with management and employees to implement targeted programs that dropped the incident rate from over 12 to less than 2.5 over an eighteen-month span."

Advice. Tell them about your individual initiative. Offer proof using real examples that you deliver more than your employer expects from someone in your job. Don't give long descriptions of situations. Focus your answer on the actions you took and the positive results you obtained.

How You Can Help the Company

When you are looking for a job, an employer will want to know what you can do to help or improve their company. Now is the time to tell them of your proven skills and knowledge that you gained at some of your other previous jobs.

Question. "Give me an example of how you can help my company."

Answer. "I will take ownership of the program such that you won't have to worry about it. I will build relationships with management and field personnel resulting in trust such that we fix issues of concern before they become problems. I will specific address the concerns you have laid out in a fashion that works both for production/manufacturing while maintaining a safe and environmentally conducive workplace."

Advice. Use an example of a significant contribution you made in your past job that impacted the bottom line. Show how this ability transfers across industries from one functional area to another.

Five-Year Goal

Employers will want to know you have drive and a sense of what your future holds for you. They will want to hire someone with a sense of purpose. Employers may ask you to describe what you see yourself doing in the years to come, whether you will be at one company or another. Telling them you see yourself doing their job may not be the best way to get hired.

Question. "Where do you want to be in five years?"

Answer. "First, I would like to see the company as one of the top SH&E performers in our business sectors. Second, I would like to see the necessary programs in place such that employees go home just as they came to work, unharmed. And thirdly, I would like to see SH&E as an integral part of the management team working closely and seamlessly with our management team."

Advice. Avoid the urge to describe job titles; this makes you seem unbending and unrealistic, since you do not know or control the system of promotion. Describe new experiences or responsibilities you'd like to add in the future that build on the current job you are applying for.

Show Your Determination

There is Good Determination and then there is Bad Determination. Good Determination is when you see a project through from the beginning to the end, no matter what obstacles you had to overcome and the result is a positive one for the company—like saving the company millions of dollars. Bad Determination is when you are blinded by your goal, the results to which can be detrimental—like stealing millions of dollars from your company.

Question. "Give me an example of your determination."

Answer. "I have spearheaded many initiatives throughout my career, but there is one I am particularly proud of. While working at one company I was challenged to reduce an incident rate that was well under control. The incident rate was 1.77, well below the industry average. After stepping back to ponder the situation, I formulated a plan, briefed management, and worked through all of the opposition convincing them of my forecasted successful results. They accepted with reservations. However, I was meticulous in implementing the program such that one year later our incident rate was 0.00. What made this even more difficult was that midway through my initiative, we were bought by a competitor. I was charged with keeping the employees focused on the program and not the acquisition. This demonstrated my determination to complete the program in the light of other distractions.

Advice. Describe your professional character, especially diligence. Describe a time you persevered to accomplish a goal. Give proof that you persevere to see important projects through, and to achieve important results. Demonstrate how you gather resources, how you predict obstacles, and how you manage stress.

Salary Requirements

You want to avoid any discussion of money if possible, at least in the first interview. That will come about in subsequent interviews, when the employer decides he like you and wants to hire you. The more certain the company is that it wants you, the higher the money. But you cannot afford to ignore the question if the interviewer raises it. A possible answer is to say that the job is more important than the money and that you would like to further discuss your capabilities and qualifications. When you do get to the salary negotiation, avoid mentioning a specific figure. If it is too high, the employer removes you from consideration. If it is too low, you have undersold yourself and perhaps even downgraded yourself in the employer's view. Telling that you might work for peanuts will get you the job, but I hope you love peanuts. Just state your salary on your last job and let the employer make an offer.

Question. "Tell me about your salary expectations."

Answer. "Current salary information published by our national association indicates a range of $60,000 to $70,000 a year. While I'm not certain how your salaries compare to the national norms for this industry, my feeling is that my value would certainly be in the upper half of this national range."

Advice. Turn the question around. Ask the interviewer first to discuss the company and the approximate pay range for the position. Then answer in general terms in line with where your qualifications fall in relation to the job requirements. Also, mention the market value for yourself, again in general terms.

Off-the-Wall Questions

Some companies will ask you stress interview questions (as if you are not stressed enough) to determine how you draw conclusions, how you react in a stressful situation, or just to annoy you. Here's an example:

Question. "Why are manhole covers round?"

Answer. "Manhole covers are round because sewer pipes are round. Sewer pipes are round because round pipes are much stronger than any other shape. Pipes placed underground must endure a tremendous amount of stress, from earthquakes to frost." (They also can be moved by one person and won't fall into the hole on top of a repair person!)

Advice. Don't mention that this topic was not the thesis statement of your paper and you have better things to think about. The interviewer is trying to see if you can think and work under stress and wants to see you sweat a little: How quickly will you come up with an answer?

Illegal Interview Questions

Illegal interview questions probe into your private life or personal background. Federal law forbids employers from discriminating against any person on the basis of sex, age, race, national origin, or religion. For instance, an interviewer may not ask you about your age or your date of birth. However, he or she may ask you if you are over 18 years of age.

If you are asked an illegal question at a job interview, keep in mind that many employers simply don't know what is legal and illegal. One strategy is to try to discern the concerns behind the question and then address them. For instance, if the employer asks you about your plans to have children, he or she may be concerned that you won't be able to fulfill the travel requirements of the position. Is it illegal? You bet. But it is to your advantage to try to alleviate his or her concerns.

Try to draw out the real issue behind the question by saying something like, "I'm not quite sure I understand what you're getting at. Would you please explain to me how this issue is relevant to the position?" Once the employer's concerns are on the table, you can reassure him or her by saying something like, "I'm very interested in developing my career. Travel is definitely not a problem for me—in fact, I enjoy it tremendously. Now, let me direct your attention to my experience and expertise in . . ."

Alternatively, you may choose to answer the question or gracefully point out that the question is illegal and decline to respond.

Calmly saying something like, "That question makes me uncomfortable. I'd really rather not answer it," will usually get you out of a tight spot without blowing your chances. Avoid reacting in a hostile fashion—remember that you can always decide later to decline the job offer.

Ask the Right Questions

Regardless of economic conditions highly qualified SH&E professionals are in high demand. The competition for the best positions is as tough as ever, so improving your interviewing technique can make a big difference in whether you're selected.

To stand out from the competition in an interview, you must thoroughly research companies that most interest you and then ask good questions. Because the right queries can dramatically boost your advantage, you should create original and probing questions.

It's always a challenge to make a meaningful impression on hiring managers. During a typical interview, the manager initiates discussion by providing basic information about the job opening and the organization. Candidates usually have less than an hour to learn what they need to make a decision about the job and a favorable impression. Using your time wisely is important if you hope to become a finalist. These suggestions can help you begin the process.

Determine in advance the type of information you want to learn during the meeting. Write down your questions. They might include, "Why do I want to work for this organization?" "How can I determine the core challenges of this business?" "What are the major barriers to the company's growth?" and "What kind of culture does this company have?"

After deciding what you need to know about the company, consider your audience. What impression do you wish to give interviewers? What are their titles and focus within the company? How busy are they? If they're pressed for time, you may have to prioritize your inquiries.

Ask questions about the organization's competitive advantage and key reasons it has done so well. You'll uncover valuable information, which will help you understand the company's recent and historical challenges.

Questions about the company's mission and the interviewers' vision of the organization and its most important accomplishments will broaden your knowledge base. This information becomes essential as you continue to interview with the firm. The more you demonstrate an understanding of the company's real business issues, the more positively you'll be perceived.

If you wish to explore cultural issues, ask about company heroes and heroines. Find out, for example, if employees are promoted to top jobs because they're entrepreneurial or good overseers. This will give you a clear sense of the organization's values and criteria for top performance and the behavior it rewards.

Keep track of the information you acquire from interviews. If you're a finalist, use it to tailor your questions to the schedule and interests of hiring executives. By showing you know the company's issues, you'll gain credibility and enhance your effectiveness in the minds of those who must ultimately choose the winning candidate.

One final point about questions. You should ask questions, but keep them to a minimum. Most people feel that asking a number of questions will prove their intelligence and expertise. However, asking numerous questions increases the risk of making potential employers feel uncomfortable or putting them on the spot. More often than not, most of your questions will be answered before you have to ask them.

Questions you may ask them include:

- What would my role be?
- Where would I fit into the organization chart?
- What is the management commitment to this function?
- What is the growth potential?
- Who will be my supervisor? Who is e his supervisor?
- One of the last questions, what are the benefits? (medical, dental, insurance, vacation, etc.)

Knowing the types of questions interviewers will ask and the proper scope of yours should enable you to avoid the common pitfalls of job interviewing through advance preparation.

Judicious Bragging is the Key to Winning the Job

Successful job hunting requires a competitive instinct. Abraham Lincoln once said, "Things may come to those who wait, but only those things left by those who hustle." You have to out-hustle five or six competitors who are looking for the same job by non-stop interviewing and by wanting the job when you have that precious 20 to 30 minutes to sell yourself in a job interview.

There are two basic reasons why many professionals fail in interviews:

- They are too shy about telling the prospective employer about their past on-the-job accomplishments. It may be work that has resulted in cost savings, new ways to increase productivity, the securing of important new business, or anything else of bottom line significance.
- They fail to impress the interviewer by giving a meaningful recitation of accomplishments. Many professionals interviewing for jobs fail to win an offer simply because they failed to sell their accomplishments.

Both of these failures relate to the competitive interview; succeeding there is the most important requirement for winning a new job. But too often, professionals approach the interview as though they are almost afraid of it. They have a tendency to behave modestly at the job interview. Although humility is usually considered an attractive trait and something we are brought up to observe from the time of childhood, it has, rightly or wrongly, no place on the job-hunting trail.

You have 20 to 30 minutes at the most to convince the employer that you are wonderful. If you don't tell him how good you are, who will? One of the hardest tasks for professionals is learning how to brag effectively, overcoming a natural tendency toward modesty without coming across as egomaniacal. The line between self-confidence and arrogance can be a thin one, so you must learn to maximize your accomplishments without antagonizing the interviewer.

To succeed, you have to be prepared in advance with a mental checklist of your most important job-related accomplishments. Then cite your past accomplishments rather than dwelling on personality characteristics.

Let the interviewer know about any praise your accomplishments won from your former supervisors. Also make a point of mentioning any awards or honors you have received in this field or in related fields.

When discussing your accomplishments, take as much credit for yourself as you honestly can. Remember that you are at the interview to sell yourself, not your former co-workers. However, never criticize a former or present employer. If you whine, you will brand yourself as a complainer who no one wants to hire.

If the interviewer should compliment you, accept it gracefully. A job interview is not the place for self-depreciation. There is only one appropriate response to a compliment: a thank-you, with a smile.

The most important element of the job interview is making the prospective employer like you. All other factors being equal, the job will go to the person the employer likes the best. You need to listen for clues from the interviewer's questions and present yourself as the person he or she wants you to be. Consciously or not, most employers tend to hire in their own self-images.

Despite the advice that is often given to ask pertinent questions, you should avoid putting the interviewer on the spot. There is always the danger your questions could be misinterpreted and cost you the job. Normally, all of your questions will be answered by the interviewer.

One thing should be stressed, the prospective employer is always right. He knows what he wants and what he is buying for his company. One of the surest ways to antagonize the interviewer and remove yourself from consideration for that job is to tell him how to run his business. Just discuss your qualifications and let the interviewer decide how you might fill the company's needs.

When the interview is finished, you should not be afraid to "ask for the order." That demonstrates the kind of aggressiveness that employers like, because it shows them you want the job. Ask the interviewer when a hiring decision will be made, and tell him you will check back with him before then. Find out if successive interviews will be taking place, and when you can expect to be called back in.

The loser in the job search is the professional who everyone likes, but who cannot get up enough nerve to ask for the job when interviewing. The individual knows what is required to succeed in a competitive employment interview, but lacks the ability or the inclination to put himself or herself over the top. The prospective employer appreciates someone who has a pleasant personality. It is one of the major pluses for a professional. But more important is the professional's drive and aggressiveness that indicates the ability to get the job done. "Super nices" leave themselves by the wayside because of their passive approach to the job interview.

You need to use every advantage you can muster, especially in today's competitive job climate. Anything less is selling yourself short. And smile, because that also helps to win a job.

The Tone of the Interview

Would you ever turn down a job based on the tone of the interview? The tone I am referring to is that you feel the interview went poorly, and perhaps revealed something about yourself or the company that did not sit well with you.

I personally have never walked away from a job, because what I do before I interview for a job is get to know what the job entails, and find out if it's something I would want to do, could do, and would enjoy doing.

If a person who's interviewing discovers that the job just isn't a fit, I recommend that he or she at that point say to the interviewer, "I'm not sure that this is the kind of job I'm looking for. Can I ask you a few questions?"

And then I would question the interviewer about the job, what it entails, what skills are involved—all the questions that I have in my mind of why I might not like this job. And I think it's worth walking away and being a little bit picky, even though you might be desperate for a job.

There are so many opportunities out there and such diversification in opportunities, that if a person does some of these things, he or she will find a job that not only pays the kind of money he or she needs, but also brings job satisfaction.

It is hard enough to find jobs today—but finding the right job can be very hard. What I do think is that there are a lot of people shooting themselves in the foot by taking the first job that comes along and then hating their work. They get locked into that job and make a career out of it, but they hate to go to work every Monday. So they have blue Mondays and thank-God-it's-Fridays, which is a terrible lifestyle. Their spouse hates to see them come home and their kids hide under the bed because this person hates their work.

Well, sure, I've walked away from a number of jobs because of the tone and my gut feeling. But what I tell job hunters is they must have six to ten things in the works. First of all, five will fall away through no fault of their own. And other ones they just need to walk away from because they're inappropriate for them.

I have not taken a job a number of times because the fit was not there with the person I'd be working for. Fit is the single most important issue.

Second is if the interviewer says, "Look, this is it. Take it or leave it." My response would be, "Leave it."

How to Close the Interview

How to close the interview is a interesting problem for numerous job seekers. Should you just thank the interviewer for the interview and walk our without further conversation? Should you ask if you will be called back for subsequent interviews? Should you ask when a decision is going to be made? And is it wise to appear to be aggressive, or is that a turnoff as far as the interviewer is concerned?

The secret to closing the interview is using the salesperson's technique: do not be afraid to "ask for the order." Tell the interviewer that you want the job and ask him or her what the next step is. That demonstrates the kind of aggressiveness companies like because it shows them that you are concerned and interested.

What you do not want to do is just fade away, walking toward the door with little or no comment. That tells the interviewer that you are not aggressive or interested enough to warrant further consideration for the job.

If you have responded well in the interview, you have probably made a favorable impression on the interviewer. You want to reinforce that with a strong closing impression. If the interviewer does not volunteer a timetable of when the decision will be made, that does not mean you are not under consideration. It may just mean he or she has not had time to figure it out yet. Your response in that case might be: "When shall I call you?" Assume that the interviewer could not possibly consider someone else.

Shake the interviewer's hand and most importantly, smile. Employers like to hire happy people. If you put on a long face, the interviewer may think that you are a grouch, not able to get along well in the workplace. Also, sometime before you leave tell the interviewer (if he or she will be your prospective boss and if you feel it is a true statement) that you look forward to working with him or her. It just sets a positive image before leaving.

One warning: keep in mind that the interview is never over until you are literally out of the parking lot. One job applicant who felt that a job offer was a cinch clicked his heels and hummed a happy tune to himself as he walked down the corridor from the room where he was interviewed. He was observed by some executives of the company, and did not get the job. Humility is everything.

Be sure to collect business cards from everyone you interview with. When you get back home write a thank-you letter and either send it via mail or email. You will be able to tell which one is preferred during the interview. This lets the prospective employer know that you are excited and ready to go to work.

How to Handle Objections During an Interview

The interviewing process is similar to an elimination tournament in sports. You compete against a rapidly-diminishing group of people, and once you lose, you're out of the competition. At each stage, the competition becomes more intense and only the best candidates make it to the final round.

While the interview is not a contest of physical strength, it is a strategic process conducted on two levels, the spoken and the unspoken, and you must communicate on both planes. Be aware of the words you choose and the way in which you deliver them. At the same time, read the interviewer's verbal and visual signals to determine how your words are being received. Do you suspect a misunderstanding? If so, stop what you are saying, and address that concern before proceeding.

In many cases, what appears to be an objection may merely be a question. For example, if an interviewer says, "Do you have an SH&E degree?" he is asking a question, and you should answer it. If you don't have a SH&E degree, follow up by asking if one is necessary for the position. If he responds by saying, "We only hire CSPs," then it's an objection.

The key is not to feel defensive or devastated, since an objection does not automatically eliminate you from the competition. Think about how you can deal with it tactfully, without allowing your emotions to enter the interview. Discuss the issue objectively, and respond to it methodically.

When to Handle Objections

There are three times when it is best to handle an objection. First, you should anticipate it and address the concern before it comes up. Second, many misunderstandings will surface during the interview unexpectedly and must be dealt with at that time. Third, you may find it more to your advantage to postpone discussing the dilemma until later.

Anticipate Objections

You may find that one particular objection comes up regularly. For example, an interviewer may think you are overqualified to work for his company. In general, an objection that you are overqualified is a smoke screen. By the time you get to the interview, the potential employer has seen your credentials and experience. This information can be deduced from your resume. Therefore, there is probably some other reason for this action. Defuse the objection by gently countering with accomplishments during your career which directly apply to the job.

Handle Objections as They Come Up

Some objections will surface during the course of the interview. If you don't discuss and remove them immediately, the interviewer may think you are avoiding the issue. That thought may linger in his mind after you leave, overpowering his good feelings about you. His intuition will tell him something is wrong, and you may be eliminated from the competition. You can handle objections effectively if you follow these steps:

1. Listen. Let the interviewer express himself completely. Don't interrupt with words or body language indicating you've heard this before or that it's of no consequence. Sometimes, as he is talking, the interviewer will either answer it himself or realize how trivial it sounds.

2. Pinpoint the real issue. Is the interviewer simply asking a question? Is there a misunderstanding, or does a real objection exist? Define the issue so you respond to the actual objection. If it's a question, answer it. If it's a misunderstanding, clarify it. If it's a true objection, go to step three.

3. Smile. Speak with confidence. Let the interviewer see that you are not going to become defensive, and he'll be less apprehensive about the objection.

4. If necessary, gather more information. As you listen to the interviewer, consider how you can rephrase his concern so it

is easier to address. If you need more time to think, respond with a question that will keep the interviewer talking

Ask a question to limit the scope of the conversation. For example, if the interviewer says, "You don't have the right experience," you should ask, "What is there about my experience that you feel is lacking, specifically?" His answer will define the parameters of the discussion.

Respond to a question with a question to give yourself more time to think. For instance, if the interviewer objects to your salary requirements, you could say, "How much did you plan to invest?" or "What budget guidelines have you established?"

5. Respond. Once you have rephrased the issue, respond to it. There is a proven technique you can use to accomplish this. The first is a formula you can remember by the acronym DARE. With this, you can either Deny the objection, Admit it, Reverse it, or Explain it. Here is how you can use each of these concepts:

   Deny it. "Actually Mr. Jones, that's not correct, I wasn't laid off from that job. In fact, here is a letter of reference written by the president of that company." Be delicate so you don't create an atmosphere of confrontation or the feeling you are out to prove the interviewer wrong.

   Admit it. "That's correct, Mr. Jones, I don't have any work experience in that area. But I do have related experience in this area. I also have a desire to learn your company's way of doing things. My lack of experience will actually be beneficial because I don't have any preconceived notions about how things should be done. Isn't that important, Mr. Jones?"

   Reverse it. Try to catch the interviewer off guard by making his objection the reason for hiring you. "That's right, Mr. Jones, I do request $5,000 more than you offered, and that's exactly the reason you should hire me." Then go on to explain how the increased cost in wages is a bargain for the experience you bring to the position.

Explain it. "The reason I didn't go to college immediately after high school is that I couldn't afford to. I needed additional income to support my family. But let me show you what I have done to continue my education while I earned the money to return to night school...."

6. Confirm agreement. Once you rephrase the objection in your terms and respond to that new definition, get the interviewer to confirm that he agrees with what you have said. It doesn't matter if your reply sounds good to you. It must convince the interviewer that his concern is no longer valid. Make sure his objection has been eliminated, and then go on to the next topic of conversation.

Put It Off Until Later

There are times when your case would be better served if you delayed handling the objection until later in the interview. For example, if you are in the midst of handling one objection and another one surfaces, you should delay the second until you have confirmed that the first has been handled. When handled properly, you can delay your response without concerning the interviewer. After you acknowledge the fact that the objection exists, tell the interviewer that by finishing your present topic you may at the same time address the issues underlying the next question.

Removing an interviewer's spoken or unspoken objection is not difficult. As you speak, watch for the interviewer's reaction to what you are saying. Listen and watch for verbal and visual signals indicating the way in which you're being received. If they are positive, reinforce that feeling. If they're negative, find out why and remove the issue. Smile, relax, and confidently participate in the discussion. Practice these skills and they'll become second nature to you. Use your common sense to tell you when to cease being aggressive, and you'll increase your interviewing success ratio dramatically.

Next Level Interviews

What often makes the corner office so elusive is the patience and determination of the interviewee. This is because you may have to endure up to three different interviews with different groups of people in the same organization. Often they have different expectations and biased views of what the role of the position in the organization will be. They are often held on three different days over a period of a month or two. You have to then figure out if you want to spend vacation on the job opportunity. You have to answer the question, "Is it really worth it?" Many times, SH&E professionals determine that it is not and the corner office remains unattained. On the other hand, what can happen is that the interview changes mid-course and other high ranking officials who were not on the original agenda are called into the interview. . You have to be prepared and study up on the company just in case you get fast-tracked. The downside is that interviews are emotionally draining. Being "on" for an extended period of time can drain the life out of you. So make sure you've had a good night's sleep before the interview.

Usually the first interview is with your prospective boss and a few of his or her peers. There is really nothing different here, other than a determination of whether you will make the cut and go on to the next round. Sometimes this is done by phone so they can narrow the field and won't have to bring in all the candidates.

There are usually two ways to go in the second interview, either to operations or upper management. If upper management is available during the first interview, you could fast-track through that hoop or fail miserably if you are unprepared. I've actually seen where a candidate is made an offer before they leave the building if he or she hits home runs here. Often, the company has been looking for a long time and hasn't found someone who can meet all their needs. When they find the perfect candidate, they strike and strike quickly. So be ready for an offer before you leave the building if this happens to you. If they want an answer right now, tell them that it sounds great but regardless, you need to sleep on it and/or speak to your spouse. Reassure them that you think it will be yes, you just want to confirm it.

If you go to the field to visit operations, find out what the dress code is and follow it. If you are going to have a plant walkthrough, dress in business casual. Be ready to ask questions about the specific operation and, more importantly, be ready to answer detailed specific questions about how you can fit in and help the organization. This is where your interpersonal and technical skills will be put to the test. Your experience and education will come to bear and how you get things done will become evident. This is a make or break opportunity, so you must be on your "A game". Be prepared and study up. This kind of stuff usually doesn't come easily, so you must work at it to be completely ready. It only takes one bad review for you to be eliminated at this point, so be on your toes and be ready to answer those operations-specific questions.

If you pass muster, then you have successfully made it to the final level. This is where you meet the CEO, the president, and maybe even the COO. Be careful, you may have already gotten the job. The goal here is to not mess up. Do your homework. Usually, the company performance information is on their website. Download it and get familiar with it. Be able to talk to specific items, such as company key performance indicators (KPIs) and projected growth. The 10K (annual) report is also downloadable from the company website, and it details all the company's performance from the previous year.

Also, print out the operations sections and the biographies of the board of directors and company officers. If there are pictures of them, print those out, too. You can use those to identify them if you see them during the interview. Find out if any went to the same college you went to. Use this to your advantage to build a bridge and to give yourself credibility. Even if they went to a rival college, you can use it to your advantage. Just be careful not to put the other college down in your discussion. Do they all know each other from previous companies? Have they been working together for years? Analyze and highlight common traits of all of them. You never know who you might meet during the interview, intentionally or by accident.

If it is a meet and greet, act the part. Discuss how excited you are about the opportunity and that you feel you can integrate well into the culture of the company. Be ready to discuss financial components regarding how you think you can save the company money. But don't overdo it. You will be able to tell if it is an interview or a meet and greet. The key is to be quiet and listen. Answer the question if asked. Be short and specific. Don't let the silence kill you. That is where most SH&E professionals stumble. Silence is okay. Just wait for the question.

Before it is over you will pretty much know how you did. Although there were times I thought I messed up, I found that I demonstrated backbone. They may then start discussing things like salary, when you are available, and benefits. It's not over yet—don't let your guard down. When you get the verbal offer, then you can start to negotiate salary, which brings us to the next section.

Six Questions Key to Evaluating a Job Offer

Once you've successfully navigated the minefield of the interview or interviews, the task is not over quite yet. You now have to evaluate the job offer. There are six questions you should ask yourself in evaluating a job offer to avoid selecting the wrong job.

Your first objective as a professional is to get the job offer, and that is where your energies should be focused, but deciding whether it is the right offer requires careful analysis. If the choice is not made carefully, it could lead to an unpleasant and unproductive job experience, as well as putting you on a career sidetrack or into a job you do not like.

1. Does the company think as I do?

One of the major considerations is whether the company has a personality similar to yours. You will be satisfied and most valuable—last longer and find the experience more beneficial—when you are able to smoothly integrate with the company and possess a comparable temperament.

2. Do I like the people I have met and what I have seen of the working environment?

This is closely related to the previous question and equally important. Often, professionals will choose to ignore the warning signals of what may be an unfavorable work environment because of a higher salary, promises for rapid advancement, or similar factors. You should weigh the people factor very carefully in evaluating the job offer.

3. Is it a high profile or low profile job?

Some professionals prefer low profile positions, but you need a high profile job if you want to be promoted. If you are at the center of corporate activity rather than confined to an isolated department or geographical area, you will have the opportunity to make contributions to the company's profitability that will be immediately noticed.

4. Would I be willing to stay on this job for at least four or five years?

You should plan on an acceptable period of tenure for each job. To a majority of companies, four to five years can be considered the right length of time today, not too short or too long (at least until you are over 45 years of age). The extremes in either direction could weaken your employment record. People who do not stay long enough on one job are frequently branded as "job-hoppers," and those who stay too long become identified as "one-company people," or individuals who do not have the ambition to handle greater responsibilities.

5. Is the compensation/benefits package what I want?

Many professionals make the mistake of lowering their salary requirement in order to become quickly re-employed, or in the belief that they can rapidly work themselves back up to their former income level. If the job does not carry the income you feel you need, it is bound to be a source of discontent that will have an adverse effect on your job performance and your sense of well-being. Forget it, unless your immediate monetary needs force you into an acceptance.

6. Is the job likely to cause any family problems?

If the position is one that may require relocation, long hours, evening work, or a heavy travel schedule, these factors should be weighed against the probable impact on your family. It is difficult to achieve and maintain a healthy work environment when there are continuing personal problems at home.

Why Professionals Fail to Get the Best Salary

Many professionals start new jobs for less money than they should be getting because they do not know how to handle salary negotiations. There are notable exceptions, such as discharged managers who have received outplacement counseling, and as a result possess the knowledge of how to handle salary negotiations. However, most professionals are so anxious to join or rejoin a payroll in today's competitive job climate that they do not take the time or thought required to properly negotiate salary and benefits, or in some cases, do not know what is required.

While the most important part of any job interview is creating a favorable impression on the employer, salary negotiation represents a close second. To obtain adequate compensation, the professional should keep the following advice in mind during job interviews:

Avoid the subject of salary and benefits during the first interview.

If these subjects are brought up too soon, the employer becomes convinced that you are more interested in yourself than you are in the company. Ideally, you should go through the first interview without mentioning money. Concentrate instead on telling the employer what a good job you can do for the company. The money gets better when the employer decides that he likes and wants you.

When you are asked how much money you want, try to throw it back to the interviewer tactfully.

Let the employer decide how much he or she thinks you are worth to the company. If you name a figure that is too high, you have priced yourself out of that job. If you name a figure that is too low, the employer will tend to think you are not competent to hold the job.

If the figure the employer names is too low, do not hesitate to tell the interviewer but first tell him or her that you want to work for the company. Then tell the person why you just cannot take the lower salary. You should have in mind how much you want and a minimum beyond which you will not go in advance, but keep this figure to yourself.

Hold to your salary requirements.

The employer may decide that he or she wants you, but also knows that many others will be applying for the same job and wants to find what the lowest salary is that you will accept. Do not compromise for less money. The job market is strong and receptive to qualified individuals, and there are other jobs that will give you what you want.

Negotiation During an Interview: Body of Evidence

In negotiation, understanding what isn't spoken can make all the difference in the world.

Reading the body language of those you negotiate with can give you a tactical edge over them. How so? A person's body language can clue you in on how you are doing during a job interview. Body language can tell you if you've been accepted, scored points, lost points, backed someone into a corner, caught them in a web of deceit, and acting rather than telling the truth. Here are six steps that can help you read what isn't spoken.

Mannerisms. Observe others' mannerisms very carefully during the first, and usually friendly, part of your meeting. If their mannerisms suddenly change later in the discussion, it could mean they're putting on an act.

Movements. Watch for exaggerated movements or extreme enthusiasm. Like poker players who hurl chips on the table or slam down cards, this usually means they're holding a weak hand.

Breathing Patterns. Pick up on your opponents' breathing patterns by watching their shoulders. Faster breathing high in the chest will make their shoulders rise and fall more than normal, and that can mean they're nervous or lying.

Distractions. Don't take the bait when your opponents slide papers across the table and ask you to read them. Instead of breaking eye contact, say, "Tell me about it. What does it say?"

Signs of Deception. Consider these as possible signs of deception when opponents use them while speaking: covering the mouth with the hands, rubbing the side of the nose, jerking the head quickly, and leaning away from you.

Positive Signs. Look for positive signs that show you can trust those you're negotiating with. The wider the gesture, the more you can trust them.

These may seem like simple ways of reading people, but it is amazing that if you keep your approaches simple, they will yield tremendous results. Don't just look at the papers in front of you, look at the people and read what they are saying in their body of evidence.

Closing the Sale

An interview is a strategic event that is part of a larger process. Generally, you'll go on two or more interviews before receiving an offer. Each one leads you closer to your ultimate objective, which is to get the job. So before you leave any interview, you should have some commitment for the next step in the process. If you depart without this commitment, you're less likely to be asked to return.

With this in mind, an interview should not just end, it should conclude. In too many instances, the interviewer stands, offers his hand, thanks you for coming in, and informs you that he will get back to you soon. Most interviewees shake hands and leave without knowing where they stand in the process, or how and when they'll hear from the interviewer.

If you want to be asked back for another interview, make arrangements to do so before you leave. Simply ask a question that will require the interviewer to commit to the next step. In the vernacular of salesmanship, it's called closing the sale.

1. Begin with a direct request for action.

As you sense the interview is coming to an end, begin to take control tactfully. Summarize all the reasons why you'll make an excellent employee, then ask a question that seeks agreement on the next step in the process. It should elicit either a yes or no response. If the answer is yes, agree upon the time and date, and leave. If it's no, handle it as an objection and try again. Don't make a statement such as, "I'd really like to have this job." That only indicates your opinion and doesn't require a response.

Instead, ask the interviewer a question such as, "Let's create a scale between one and ten to define where I stand as a candidate. One is the lowest score, indicating I won't be given further consideration. Ten is the point at which you'll cancel all the other interviews and offer me the job. Given the information we've discussed today, where would you rank me on that scale?"

Most interviewers won't commit themselves too early in the process and will probably give you an average score of five or six. If it's below four, it may be in your best interest to pursue employment elsewhere if you can't increase your score. If it's close to ten, ask when you can start. But if your rating is in the mid-range, ask what you need to do to improve it.

2. Respond indirectly to neutral or negative responses.

If your ranking is low on this scale, find out why. He may say he doesn't have enough information upon which to base his decision. Determine the areas in which he would like to know more, and then fill in the missing details. Follow with a clarifying question: "Is that all you needed to know about that subject?" and then ask another direct question to get commitment.

If the interviewer still seems to be hesitant, use indirect techniques to find out why. Here are several methods for bringing the interviewer's concern to the surface diplomatically.

The Switch. When using this method, you respond to a negative answer with another question.

Interviewer: I think you'd probably make a good employee, but I'm just not sure you have enough experience.

You: If I can demonstrate to your satisfaction that I can perform in this position successfully, will you hire me?

The Negative Yes. Another technique is particularly useful when the interviewer is not willing to divulge his true objection. You simply ask a question, the answer to which is most likely to be "no." But every time the interviewer says "no," his response can be taken as a "yes" if the question is phrased properly.

For example, on the first interview your objective may be to schedule a second interview. Your closing statement should be phrased to get a definite time and date to return. If the person will not commit to a second interview, use your skills to find out why he won't ask you to come back:

You: "Mr. Jones, I feel good about our discussion today, and I am confident that I could make a valuable contribution to the company's goals. I understand that you have several other people to interview, but I would like to schedule another time to see you. Would 9:00 a.m. next Tuesday fit into your schedule, or would 11:00 be better?"

Interviewer: "No. That won't be necessary. I'll think about the topics we discussed today and call you in a few days."

You: "I can appreciate that, Mr. Jones. But just to clarify my thinking, what is it that you want to think over? Is it anything to do with my education? ("No."). Does it have anything to do with my summer job at the factory? ("No."). Then does it have anything to do with the salary I have requested? ("Well, it is a little higher than I had wanted to pay for this position.")

Then proceed to handle it as you would any other objection regarding your salary by convincing him in some other way that you are worth what you are asking. You'll have to decide when to stop pressing for commitment. If you sense the interviewer is becoming annoyed at your insistence, back off and accept the situation as it stands. Demonstrate not only your persistence and interest in the position, but also your common sense and intuitive skills.

Don't try to be someone you're not. You must start with an objective and take action that will lead you to your goal. As you speak, watch for the interviewer's reaction to what you are saying. Listen for verbal and vocal signals indicating the way in which you're being received. Then once you sense the interviewer coming to a conclusion, begin your closing strategy. Summarize all the areas of agreement and ask a question that will reach your objective.

Smile, relax, and confidently participate in the discussion. Attend every interview with an open mind and a solid understanding of your skills and accomplishments. Practice these skills and they'll become second nature to you. Use your common sense to tell you when to cease being aggressive, and you'll increase your interviewing success ratio dramatically.

Summary

As you can see, getting to the corner office is a combination of skill, timing, and being prepared. Your resume needs to be fit for the purpose. Your cover letter needs to let them know that you know the job, you can do what the requirements are asking you to do, and that you are excited about the opportunity. You need to be ready for many different interviewers with different backgrounds and biases that come to bear on your position.

You must be ready for all kinds of questions, so you'd better know yourself, the company, and have an understanding of the people you are speaking with. Be ready to interview with a line employee all the up to the CEO. If there are multiple levels of the interview, you need to be patient and persevere so that you do everything you can to be the one who comes out on top.

Be on your toes during the interview and ask some pertinent questions that reflect you were listening and not letting your mind wander. Also, you had some hard questions that could be asked. These are just some examples. You will need to come up with some responses that fit who you are. It is easy to see right through someone who is faking it, so be true to yourself.

Sometimes there are objections during an interview: be ready to respond. Anticipate them and be ready to respond honestly and truthfully. Look for visual and verbal clues and try to mitigate them if possible.

Lastly, the most difficult thing for SH&E professionals is negotiating a salary. Develop your game plan and execute it step-by-step. Determine at every juncture how you will proceed to maximize the final offer you can get without making them think you are greedy. Bringing all this together, you can close the sale in a win-win situation where everyone is happy and you can go to work for the company making the salary you are happy with and the one they are happy offering.

Success Tips

Smooth Sailing Through the Seven Cs

1. Creative. Think outside the box to address job requirements.
2. Credible. Your accomplishments must not be exaggerated.
3. Convincing. Use action words to communicate your accomplishments.
4. Complete. State enough of your background without it being a biography.
5. Current. Everything must be up to date, including jobs, publications, and education.
6. Clear. Avoid being too wordy. Say what you have to say in as few words as possible.
7. Concise. Draw a direct line between their needs and your skills.

Interviewing Tips

1. Research the company before the interview.
2. Use your network to assess the company culture.
3. Let the interviewer drive the pace of the interview.

4. Be on time to the interview.
5. Keep a positive frame of mind.
6. Arrive at the interview alone.

Avoid the Big Mistakes

1. Failure to look the interviewer in the eye
2. A limp handshake
3. Arriving late
4. Asking too few questions
5. Intimidating the interviewer
6. Poor oral skills
7. Lack of career planning
8. Making excuses, evasive answers
9. Lack of maturity
10. Cynical attitude
11. Lazy

Interviewing Etiquette

Follow the interviewer's lead.

Meal Etiquette

Understand that you are not there to eat but answer questions.

Closing the Interview

1. Ask for the order. "I would like working for you," or "what is the next step?"
2. It isn't over until you are—literally—out of the parking lot.

Next Level Interviews

1. Print out officer information. Read them. Note any positives for you.

2. Read the annual report. Tag things that stand out.
3. Wait for questions.

Six Questions Key to Evaluating a Job Offer

1. Does the company think as I do?
2. Do I like the people I have met and what I have seen of the working environment?
3. Would I be willing to stay on this job for four or five years?
4. Is the compensation/benefits package what I want?
5. Is the job likely to cause any family problems?

Why Professionals Fail to Get the Best Salary

1. Avoid the subject of salary and benefits during the first interview.
2. When you are asked how much money you want, try to throw it back to the interviewer tactfully.
3. If the figure the employer names is too low, do not hesitate to tell the interviewer but first tell him or her that you want to work for the company.
4. Hold to your salary requirements.

Corporate Survival

Is It Time to Jump Ship?

After several years with a company you get into a rut, begin experiencing broken promises, or you find that the job and/or the company is not really going in a direction that suits you. You may also find that competitors are emerging and that the company isn't responding to them. Business is beginning to disappear. The company profits are declining. You begin to ponder, "What should I do? Should I stick around or do I jump ship?"

Along these lines, I made an agreement with the president of the company that, if I could achieve an incident rate of 2.5 with no lost time accidents, I would be promoted to a newly-created corporation SH&E manager position. I provided status reports to the president monthly, with newsletters indicating our reductions based on activities and programs implemented throughout the performance period. One and a half years into the agreement, I had achieved the goal and was just waiting for the two-year mark for the company to clear the lost time accident hurdle. About three months from the milestone, I inquired through the ranks as to the status of receiving the reward for achieving the mutually agreed-upon goal. The response was, "What agreement?"

I began to look around and quietly put my resume on the street, in addition to responding to monthly inquiries from headhunters that I had previously turned down. Within only a few weeks I had an offer that I thought would work. I gave a two-week notice and all of a sudden the president woke up. He offered me the position we had agreed on and several other allurements. I ultimately turned it down, mainly because of trust. He had lied to me once. I figured he would do it again.

The reason I wanted the promotion based on my performance was, I thought that I had mastered the position and done all that I could do for a company of this size. My job satisfaction had dipped significantly and I had become, well, bored.

Take a Risk

Making a career change is always difficult, but it's usually toughest when you have a job. The risk of rocking the boat seems enormous even when work isn't going well. The reason is simple: A job in hand, no matter how mediocre, is at least predictable. This is especially true if you haven't planned a career alternative that will allow you to leave.

I have found that people often hold on to their jobs much longer than they should. I encourage colleagues to take a much more active approach in their career management. As soon as you sense something is wrong, get help. Seek out a mentor within the organization or a savvy businessperson whom you respect outside your organization. Then update your resume (I always keep mine updated), identify potential employers and start getting referrals from friends and associates. Hope for the best at work, but don't assume that the situation will work itself out.

Perhaps the most critical aspect of making a job change is believing that you can do it successfully. Part of knowing when to leave is creating options for yourself, such as building a financial reserve, updating your skills, and cultivating an attitude that allows risk-taking. If you have no options, knowing when to move doesn't matter, because you're going to have to stay right there until somebody fires you and you have to leave, or the stress of the job gets to be too much.

Time to Leave

To determine whether you should consider jumping ship, review the following three situations every year (or more often to see if one or more describe your current status. If they do, and you believe the situation can't be corrected, it's time to move on.

You no longer believe in what you're doing or in your company's mission.

On the surface, changing jobs because you don't feel that you're contributing meaningfully to society may sound corny, but it's not. We all need to have confidence in the value and usefulness of our lives. People who believe in their work tend to be excited about it. They seek out knowledge related to their jobs and tend to perform well. Others, in turn, are attracted to their enthusiasm and seek their advice. Inevitably, these factors contribute to career success.

If you believe in your company's mission but your job is no longer satisfying, consider looking for another position within your company. Transferring to another department often is the easiest, most effective way to rejuvenate your career.

If you question your employer's integrity, you have few choices. I once worked for a company that severely compromised SH&E for the almighty dollar. Fortunately, I discovered this early and was able to return to my previous employer. I was once told by an old sage SH&E professional, "Your reputation is only as good as your employer, so pick your employers carefully."

You're no longer growing and learning in your job.

Many people believe that to grow in their careers they must move upward. But now, when downsizing and flattened organizational structures are the rule, climbing the ladder with your current employer may be virtually impossible. If you can do your job with your eyes closed, consider trying to enrich your current position. If it can't be made more challenging, it's time to find another job, perhaps by moving laterally to a position that offers different responsibilities, but with a similar pay scale and title.

To enrich your current job, volunteer to serve on cross-functional teams that will expose you to other areas of the company, such as ergonomics, reliability, or use of computer technology, or tackle problems that no one else either has time for or have failed at. This can identify you as a performer who succeeds where others have failed. Volunteering for special assignments within your organization also can lead to a better position.

You suspect that your job will be eliminated or relocated.

It's far better to take the initiative and move to a different position than passively hope that someone in the company will find another job for you. Don't ignore the following warning signs that your position could be eliminated:

- Your division's product line is being discontinued.
- You're left out of meetings you should be attending and don't receive important memos or emails.
- Your work suddenly is being criticized unfairly.
- Your boss withdraws critical support you need to do your job.

Remember, you don't need a liquid crystal display to read the handwriting on the wall. Most SH&E professionals are adept at reading unspoken signs when doing incident investigations. Apply the same to your job in these times.

I've seen many people who heard the drumbeat but didn't have enough foresight to take control of their careers. They didn't want to hear the rumors or see the signs. They blinded themselves. They were hoping that the currents would change, or that the boss would change, or that this bad month or year they were having was a fluke, rather than a trend. Unfortunately, it's significantly more difficult to find an opportunity when you're unemployed.

If you hear news that threatens your job, get your credentials in order and get as many irons in the fire as possible. Truly versatile employees rarely lose their jobs, they usually land on their feet. So even if your position is being reduced, you should try as quickly as possible to broaden your experience and seek opportunities within your organization that can enhance your career.

If you discover that your company is planning to relocate and you can't or won't tag along, you may be able to negotiate outplacement assistance. More likely, when you refuse to move you'll be on your own. In either case, you'll want to activate your network as quickly as possible and call on company colleagues for references before they move away.

Planning Your Career

There is only one person who cares about your career–it's the one reading this book. Never assume that the employer will take care of you through thick and thin. If they do, it's nice, but don't count on it. Through the years I have assembled some tried and true principles for career planning. They are:

- Never rest on your laurels, continually pursue new technologies and opportunities.
- No one will plan your future for you.
- Set a five-year plan; that is, write it down and evaluate it every year as to where you are.
- If possible, identify several paths to follow, for example, private industry versus public sector, insurance versus industry, or academic versus consulting.
- Diversify wherever possible.
- Get as many professional credentials as possible that relate to your field.
- Publish, publish, publish–but only when you have an "Aha!" idea, not when you have a "So What?".
- Present at, local, regional, and national technical conferences.
- Develop ideas for books that no one else has talked about before.

Don't Let Your Dream Job Become Your Worst Nightmare

In the beginning, job descriptions were simple and to the point. Rarely were they more than a few sentences. Here is an example for the Pony Express:

> **WANTED:** Young, skinny, wiry fellows, not over 18. Must be expert riders, willing to risk death daily. Orphans preferred. Wages $25 per week.

In 1859, Wild Bill Hickok answered this help-wanted ad for the Pony Express, confident that he was qualified to deliver the mail on time without getting killed. By reading the notice, this famous icon of the American West gained a clear, honest description of the job's requirements. That allowed him to accurately judge his suitability for the work.

Jobs today may have the same deadline pressures as a Pony Express rider's, but if so, that's where the similarities end. Positions today are now so complex and interrelated that they seem to continually change. Remember when the top job was safety? Then it became safety and health, later it became environmental, safety and health. Even later it became quality, safety, health and environmental. Now they're adding risk management and security. Some are adding ethics and compliance. And the list continues to grow. Thus, the description SH&E applicants receive can be far different from the job they ultimately perform. But since the initial description is all employers have to go on when screening SH&E applicants, they may wind up hiring a Wild Bill Hickok to do General Patton's job.

With the diverse economy, expanding professional needs, and shortages of qualified talent across the board, employers are desperate to find and keep all kinds of skilled professionals, including SH&E.

But caveat emptor. If you're offered a great new position or long-overdue promotion, my advice to you is *don't* immediately accept. If you overestimate your abilities or the complexity of the job, your dream job could become your worst nightmare. You may end up living out the Peter Principle (being promoted to your level of incompetence) right before your very eyes.

Blinded by Ego

Consider Joe Safety in the XYZ local manufacturing facility. He has been the safety manager for five years with a good track record, managing a department with two professionals. ABC Company offers him the corporate safety position with fifteen facilities scattered throughout a five-state region. The pay is great and the chance to get hard-to-attain corporate management experience is bountiful.

The job offered so much to his career that he felt he'd be crazy to turn it down, thinking that he'd pick up on the management and personnel issues simply by being *in the environment*. He accepts the offer.

Soon after starting, he realized he had bitten off more than he could chew. As he battled with management, dealt with difficult personnel and scheduling issues, and coped with state, local, and federal OSHA and EPA regulations, his confidence plunged. In his small local manufacturing facility all he needed to know was one state's regulations and deal with a few professionals and one plant manager. He had not practiced the skills necessary to be successful in his new position.

He was consistently working 14-hour-plus days to keep up with the demands. Often, he felt on the brink of a breakdown. He walked into the job feeling untouchable and figured that whatever he didn't know wasn't important in the whole scheme of things. It was far more responsibility than he was ready for.

He continued for almost two years before resigning out of concern that his performance might damage his professional reputation. Even then, the decision was difficult.

Looking back, the allure of money and professional growth as well as professional egotism outweighed common sense when the job was offered. If he had looked past his ego and really thought about the job, he never would have taken it.

High Expectations

You can find yourself elbow-deep in alligators in a new position for other reasons. For instance, in today's competitive working environment, employers tend to have unrealistic expectations of new hires, especially if you were hired through a headhunter. If you are not panning out they can get rid of you and not pay the headhunter. Furthermore, unless you check the requirements thoroughly, you may accept a job that seems doable only to find the employer expects too much of you.

In companies where employees are expected to be virtually autonomous and who wear multiple hats, this is a particularly prevalent problem. To gain more time for a personal life, some professionals may choose to resign from their corporate jobs at a large company for a production level safety position at a small facility. In this hypothetical example, after starting the job, Joe Safety learned that the resources-strapped company wanted more from him than he was able to give.

Since he lacked regional experience, Joe Safety was in the office 12 to 14 hours daily struggling to keep the safety program together. He was in emergency mode almost every minute of the day. It was exactly the opposite of what he had anticipated. Eventually, he would burn out and become completely ineffective.

Avoiding the Problem

To avoid a bad fit, SH&E professionals need to interview the company to ensure that it is a good fit, not just a better-paying job. During the interview process you need to ask probing questions. There should be a two-way interview process: the company interviews you and you interview the company. You can ask the company to take a look at the potential fit. If this occurs before you accept a job, it can head off problems from the start. To more accurately assess whether you're on the verge of accepting a potentially disastrous job, consider the following.

Examine yourself and your talents. A simple but often overlooked step in avoiding a nightmare job is to evaluate how well you fit the role. Honestly determine your strengths and weaknesses. Knowing your work style and personality will help you pass on jobs in which you won't succeed. For instance, if your ideal job is a management position and you've never been comfortable leading people, think twice. You could learn to manage, but will you enjoy or be good at it?

Performing a new job effectively may require skills, experience, or education that you lack. Is there sufficient time to prepare yourself? If you learn as you go, your new boss and company may not be pleased with your inevitable mistakes, and neither will you.

Examine the job. If you and your dream job are a good fit, study the job itself. Ask for a written job description that states the official requirements and expectations for the position. Be especially cautious if the written description doesn't match what you were told in interviews.

If the description is consistent with what a prospective manager or human resources indicated, request an on-site visit. Ask coworkers, managers, and other employees for feedback on the job and company. If significant differences arise between what you were told and what employees say, you could be stepping into a potential land mine. You'll be the loser, not your boss, if the job description doesn't agree with your expectations.

Also, understand your manager's performance expectations before accepting. If you'll need to reach specific goals, be sure they're realistic.

Seek peace of mind. Your ultimate goal should be peace of mind. Even if your talents seem to mesh with company's requirements and the expectations are realistic, you may not be able to live with yourself if the job or company violates your ethical, moral, or professional standards.

Learn about your predecessors. Were they fired or demoted, and if so, why? If you question the integrity of the job or company, no amount of prestige, challenge, or money will repair the problem. Any job that forces you to choose between company goals and your reputation isn't worth having.

Avoid the dangling carrot. A dream job may come with a big paycheck, an office offering great views, a generous expense account, and other perks. Don't be blinded by these lures. They're designed not only to attract good candidates, but also to sometimes obscure a job's drawbacks.

Separate the perks from the job, then evaluate the position based only on its performance components. Is it really worth it? If the job requires you to be General Patton instead of Wild Bill Hickok, you may not like the outcome.

Higher compensation and benefits are the norm, but some employers use them to make candidates believe an undesirable position is really a great opportunity. A trap that contains the right bait will always catch something. Don't be duped. You don't want to chew your leg off to get out of a bear trap that offered more money than you could refuse.

Some Final Tips

Don't let the thought of a dream job take control of your life if there are hints of it becoming a nightmare. If you are a Wild Bill Hickok, don't sign up for a General Patton job. Don't let your ego write a check that your abilities can't cash. As Warren Buffet once said, "It takes twenty years to build a reputation and twenty seconds to ruin it."

Make sure that the expectations of the job don't guarantee failure. If your background does not closely match the job requirements you will be in for a prolonged learning curve and in danger of imminent failure.

Talk to as many people as possible—predecessors, future co-workers, and departments you will be required to work with—to get a clear picture of the whole job. If it doesn't match your gut feelings, look at the job carefully.

If the salary and benefits seem too good to be true, be careful, there may be underlying reasons.

In the end, taking these issues under consideration may allow you to avoid letting your dream job become your worst nightmare.

Surviving Today's Corporate Downsizing

For many employers, defining the layoff line of scrimmage is a tough call. The quality of who stays and who goes is critical to corporate survival. To remain in the game, SH&E professionals must keep their eyes and ears open and their skills sharp.

When it comes to corporate downsizing, the real issue is survival. However, survival strategies are different, depending on who you are in the organization. Human resources has a twofold role:

- To ease the separation process of those employees leaving the company, and
- To assist remaining employees in the rightsizing transition period.

Here are some suggestions and background for ensuring that you have the best shot at surviving corporate downsizing.

Any company considering downsizing as a solution to its strategic concerns should first determine what its strategic concerns are and then fit downsizing into that context. Otherwise, downsizing may not provide that anticipated boost to the bottom line. No employer can guarantee its employees eternal protection in an increasingly unpredictable business environment. Even if employers did, why would anyone believe them? In the 1980s, companies acquired smaller companies, diversifying into somewhat-related industries. In the 1990s, companies divested these acquired companies to focus on core businesses because these acquisitions took time away from profit centers.

SH&E professionals can ready themselves to survive the effects of layoffs by staying on top of career options and industry trends. If you are monitoring the trends and you see negative indicators of company health, act now. If you wait until layoffs start to occur, you may have waited too long. The time to be looking is every day you go into work, every day you have a job. If you want to be a survivor you must know what the direction of the company is by fully digesting the annual report. The first page tells the mission and where you are headed.

Also, you must read reputable trade publications, or the *Wall Street Journal* or *Business Week*, on a regular basis. It is possible to find out things there that your everyday work assignment doesn't expose you to. When there is a new strategy you've got to assess its impact on SH&E so that you can be prepared for any changes that may affect you. Survival is a matter of being flexible, being willing to take risks, and staying open to various opportunities.

Approaches to Rightsizing

According to a factor analysis based upon the American Management Association (AMA) survey on work force reductions, there are certain patterns to corporate resizing: Preventionist, People Pushers, and Parachute Packers. SH&E professionals can better survive a wave of cutbacks by understanding how employers choose a corporate resizing pattern.

- **Preventionists**. Preventionists are characterized by minimizing layoffs when there is a decrease in work volume. Preventionists are more likely to be in the manufacturing industry, rather than in the service industry, and tend to be medium size. Their tactics of job sharing, mandating pay cuts, or pay freezes are more commonly used with skilled blue-collar workers and skilled white-collar workers such as programmers or technicians.
- **People Pushers**. People Pushers try to push people out of surplus jobs. These firms are not just manufacturers but are also service providers, such as utility companies. The major reason for People Pushing is technological change. To

encourage people to leave, these firms offer early retirement, voluntary separation incentives, and outplacement assistance.

- **Parachute Packers**. According to Parachute Packers, the main reason for resizing is not business downturn or automation, but the need to use staff more efficiently. This group appears in all industries.

No matter what the motive, downsizing still leaves the at-risk employee adrift in a sea of uncertainty. Natural attrition, voluntary retirements, and creative programs that allow employees to leave a company while still remaining connected to it through new business arrangements are ways employers can contain the trauma while using downsizing to its strategic advantage. Unfortunately, remaining employees may feel betrayed by the prospect of continual layoffs.

Downsizing to Cut Costs

Whether your company is a Preventionist, People Pusher, or Parachute Packer, the primary motivation to downsize is to cut costs. Often, however, companies do not realize these savings as they were initially projected. Successful companies reduce costs by reducing useless layers of work rather than blindly cutting people. The question employers must ask is: Can the company afford downsizing programs? Job retraining, incentives to leave, and severance or outplacement packages are the most expensive downsizing programs for an employer. Job sharing, pay cuts, short work weeks, and other strategies usually cost less.

Some companies determine that an incentive plan to eliminate a large number of managerial jobs that cost millions of dollars can save the company multi-million dollars a year. However, while these companies experience significant cost savings by offering early retirements, other businesses have found that they lose workers they want to keep and must pay large sums of money to lure them back.

Meanwhile, for a company choosing to reduce its work force through involuntary layoffs and dismissals, the cost can be much less than for a voluntary separation plan. However, this may be the case only if the employer can successfully defend its decisions in court on the basis that these decisions do not violate federal job discrimination laws.

Despite their hefty prices, one advantage to voluntary reduction programs and early retirement packages is a reduction in the risk of lawsuits filed by angry, dissatisfied employees. Such voluntary reductions, therefore, may save an employer millions of dollars in legal fees and lost production.

In addition to those costs directly related to downsizing programs, employers also consider indirect costs. For example, companies carefully scrutinize the effects of massive layoffs on corporate stock values. When asked if company layoffs are an indication of financial instability for investors to be wary of, employers reply that it could initially be interpreted that way, but it could also be interpreted as an aggressive, proactive management maneuver. Such layoffs could also demonstrate that quality efforts are in place to bring about productivity improvement. However, layoffs may be a sign of increasing shareholder value rather than decreasing value because you're stepping up to the need to size correctly.

The question to employees is, if layoffs are not an indicator of troubles ahead to investors, should employees consider offers of voluntary separation as the first indicator of future involuntary layoffs? The answer to this question in part is that in large companies with a nationwide voluntary program, the light probably should go on for employees as they witness some of the technological changes. Secondly, the light probably should go on for employees as they witness changes in the business market. This will translate to restructuring, and hopefully to some quality and process improvements.

Anyone who learns about quality will be invaluable to many employers. This process happens to be based on teamwork, analytical processes, and continuous improvement. Allying with the quality process or integrating with it, helps ensure survival in times of downsizing. This makes you that much stronger and more marketable.

Although SH&E professionals may accept the challenge to make themselves more marketable in today's environment of corporate upheaval, there is usually little hope of a safe haven anywhere in a corporation. Certainly, if you can demonstrate consistent results that meet business needs, you can illustrate that you are contributing to your company's bottom line. However saying that all employees in SH&E are safe is inaccurate. Layoffs are fair game everywhere. With the advent of a number of quality improvement processes, the environment must change.

Getting Prepared

SH&E professionals in jeopardy of losing their jobs can no longer rely on traditional job search techniques such as headhunters, newspaper ads, or outplacement centers. In fact, the majority of jobs never reach the newspapers. And corporate outplacement counseling sessions may only run from four hours to four weeks or more, depending on what companies can afford. SH&E professionals therefore, must turn to organizations (such as the American Society of Safety Engineers), online networking sites (such as LinkedIn, Facebook, and Twitter), churches, schools, and other community and professional associates for more extensive career networks.

In these uncertain economic times, SH&E professionals, both those victims of layoffs and those remaining, should also take advantage of offers by employers for training. Such development may make you more marketable for other positions within the organization or at other companies. For example, to handle the effects of one of its major downsizings, Eastman Kodak established a $100 million career resource center in Rochester, NY.

Another effective career option for many SH&E professionals today is independent consulting to the company they just separated from. Such consulting options may tide an employee over until a more permanent position at another company becomes available.

Being prepared may be an SH&E professional's best defense against the turbulent effects of corporate downsizing. Getting prepared, therefore, should be a goal for SH&E professionals. To succeed at this goal, SH&E professionals must be prepared to communicate effectively with employees to reduce the risk of unwarranted rumors. SH&E professionals must be prepared to reassure the remaining employees that the company values their contributions and will keep them informed as more information unfolds.

Savvy Advice on Surviving a Corporate Merger

Just when you thought the joint ventures, mergers, and acquisitions were over, they have resurfaced. Judging from recent activity, organizations in such industries as health care, oil and gas, telecommunications, consulting, and accounting believe pairing off is the best way to grow and tap new markets.

But when companies join forces, every employee, from the CEO to the new hire, is affected. Even if you have many years of experience, it is difficult to prepare for the ensuing change, uncertainty, and confusion that follows the corporate transaction. Employees must deal with fears about job security and retirement, insurance, and other benefits while phasing in changes required by the new owner and operator. Employee workloads increase along with personal frustrations. Sometimes the hardest part is the initial uncertainty and confusion.

The greatest anxiety that tops the list during these activities is clearly job loss. With mergers apparently increasing, the road will be a rough one. If your company is merging or being acquired, don't expect the transition to be smooth. You'll have to contend with organizational chaos and your own emotions. To embrace a new company and take ownership is difficult when no guidelines exist. To survive, chart your career in the new company and handle relationships with coworkers and colleagues effectively. Here are some tips.

Develop a plan. As an engineer, I talk a lot about plans. While many SH&E professionals do not have plans of any kind, you'll weather the merger transition better if you have a strategy for managing your career.

The plan should include goals for long-term career success and short-term job survival. Develop a list of networking contacts, explore different career options, and set funds aside for a financial emergency. You should also practice following up with contacts and headhunters about potential job opportunities.

Developing an action plan to guide you through the ongoing process of mergers is imperative. It keeps you from becoming confused and overwhelmed.

Maintain your professional integrity. Knowing how to satisfy your current employer while preparing to work for another is an acquired skill. Make decisions carefully to avoid jeopardizing your relationship with either regime. The layers and layers of new management may complicate even simple choices.

Double-check the effect of decisions where results aren't immediately apparent. Show the new management team that you're a consummate manager by remaining optimistic and open to new challenges and responsibilities. If you are the pessimist who believes that the glass is not just half empty but upside down, your half-life with the new management will be even shorter.

Be tactful with peers. Mergers test work relationships, especially those with colleagues whose jobs may be phased out. Your dealings with others will take on a new dimension as job descriptions change and layoffs are announced. Your position may survive, but a colleague's might not. Listen and be empathetic to his or her plight. Offer encouragement and suggestions. Minimize your feeling of relief so you don't seem oblivious to others' pain and worry. In other words, be sensitive to their pain.

Watch what you say about the merger. You'll receive subtle and pointed questions about your future from coworkers, friends, and family. Decide how to respond in advance. Know what information you can share and don't divulge details that breach confidentiality. Zip your lip when asked details that are proprietary. If you don't, and are found out, you may move from the keeper list to the lay-off list. State confidently that you hope to be in a new routine soon. Don't make negative or cynical comments about the acquiring company, since judgmental statements may hurt you in the future.

Learn to be a chameleon. If your company is acquired, you must shed your old identity and loyalty and adopt new ones immediately. Avoid the starter phrase, "When I was at such and such a company we did it this way." Nobody cares and the former company doesn't exist anymore, anyway. So, remove that phrase from your repertory. The only constant in today's workplace is change, and dwelling on the past instead of the current reality will make the transition more difficult. Remember, you can respect your previous employer and still develop loyalty to a new owner.

Assume personal control of your career path. You may be worried about a job or position change, but you don't have to feel demoralized. Be mentally tough and make positive things happen by taking advantage of new opportunities that come from the realignment. Participate in activities that allow you to feel a sense of accomplishment. I look at all of these activities as opportunities to excel and move up the food chain rather than down the food chain in the organization chart.

Manage your emotions. Unexpected changes in work environments can throw you emotionally. In time you may lose your position, employer, and colleagues. If this happens, don't minimize the loss. Allow yourself time to grieve, then move on. If you don't, you can easily harden your heart to others' plight and lose your sensitivity to other people. Take a positive attitude. View change as a stepping stone to something better. Don't dwell on your loss or the past. Assume the future offers even better possibilities.

Relieve frustration with stimulating physical activities. Sports such as tennis or racquetball, or some other sort of exercise, coupled with a healthy diet, are stress reducers. By improving your physical fitness, you'll have the stamina to cope with change and remain productive amid the turmoil.

Assume a positive leadership role. You won't score points by making difficult situations worse or by creating problems. Demonstrate leadership skills by helping employees pull together for the new team. Don't take sides in office gossip, spread misinformation, or behave like a victim, which demoralizes others and reflects poorly on you.

Realize that some things never change. Mission statements and job descriptions can be rewritten, but what's required of a good employee remains constant. Corporations need capable professionals to handle challenges, save time, and improve efficiency. If you're interested in being a survivor, show your new employer your willingness to help. Above all, don't panic during a realignment. I know that this is easy to say but difficult to do. You may have to rely strongly on your spouse, friends, relatives, and significant others to achieve this.

Find new rules for success. Not long ago, experience, expertise, and efficiency were key requirements for advancement. Today's requirements include innovation, creativity, and resilience. Develop team leadership abilities and self-reliance and be ready to act on new opportunities. Taking advantage of promotion opportunities that result from a merger may give you an edge and help you remain employed. If you assume change is bad, you may make it a self-fulfilling prophecy.

Keep events in perspective and expect some unpleasantness. Periodically weigh the pros and cons of your job. If negatives outweigh positives after the merger, it may be time to rethink your career. Explore all your options before deciding to leave, and don't be discouraged by the prospect of job hunting. Being asked to leave isn't necessarily fatal. There's always a market for talented people.

Accelerate your networking activities. Beyond professional society networks, helping peers is as important as gaining their assistance. It can be amazing how many people can come to your assistance if you helped them in time of need. The outpouring of assistance can even be overwhelming. Many employees will lose jobs because of overlapping job descriptions and new priorities, not because they're incompetent. Be aware of openings and refer others whenever possible. By being helpful, you'll receive assistance when you need it.

Create a survival kit. Determine what you must do to remain secure and calm during this period. Participate in outside activities that help you feel anchored and self-confident. Develop a support system and postpone difficult decisions in other areas of your life until your career stabilizes. Monitor your self-esteem and avoid those who trigger negative feelings. It is difficult to get a job when you interview like an abused dog. Keep your chin up and persevere. Seek to be interdependent. Your survival kit should include people who can encourage you and be uplifting and supportive.

Be entrepreneurial about your career. Explore and evaluate different options. Polish your technological skills and develop a backup plan. Daydream about ways to improve your life. If you can visualize it, you might be able to accomplish it.

Become philosophical. William Faulkner wrote that "...man will not merely endure: he will prevail." This is true in the turbulent corporate world. Even if your department is being eliminated and your future is uncertain, realize you have the strength to solve problems, create new opportunities, and above all, to prevail.

Surviving a merger can be very challenging. Develop a plan with several options. As you go through the merger, make decisions that assist you down your chosen path. Regardless, maintain your integrity and be tactful during the change. It is better to purse your lips than speak your mind. Let others talk, you listen. Try not to show your true colors regarding the merger if it does not agree with your career desires. Leave them wondering. If you have control of your career look at it as an opportunity. If you choose to leave, activate your professional network to scour the landscape for opportunities. The goal is to survive the merger, one way or another.

Dealing With a Difficult Boss

As many people in industry today well know, the comic strip *Dilbert* skewers corporate life brilliantly. Dilbert's self-impressed boss doesn't have a clue. He often demands ridiculous projects and kills sensible ones. Oh, I'm sure nothing like that ever happens at your company, but many people in the corporate world think it hits darn close to home.

Such managers are comical indeed–until one of them stands in the way of one of your major projects. That's when you begin to believe in humor, and your life is part of the manager's latest sitcom. Surprisingly, lost-in-the-fog managers are a relatively new phenomenon. Sure, there have always been incompetent managers, just as there have always been inept plumbers, doctors, drivers, and the list goes on. But a dramatic thing began happening a decade or so ago.

Until that point, a person with many years of experience tended to know just about everything in his or her particular field. The pace of change was manageable. You could even stay on top of fields where change was ongoing, such as accounting, law, medicine, and safety, health, and environmental.

This was even true about computer technology until the mid-1980s. Think back to when you bought your first PC. There was a lot to learn back then, but a brilliant mortal like you was able to know it all. Today, with the rapid change in the Internet, proliferating computer technology, not to mention regulations and governmental requirements, it requires much more work to keep abreast.

No one said work was going to be fun all the time. Fun is sometimes a function of you, and sometimes of your boss. Every now and then it happens—your boss leaves or retires and the person replacing your boss is a difficult person like Dilbert's pointy-haired boss. The end result is you have to deal with a difficult boss.

Whenever someone changes jobs I always tell them, "Pick your bosses well." If you don't, you could be in for a lot of grief. But sometimes you don't get to pick your boss. If this happens, what do you do?

Have faith in yourself. Difficult bosses rarely give feedback, and when they do it is negative bordering on castigating. You walk away feeling that you are incompetent. You have to believe in yourself that you can do a good job. After all, your boss isn't going to tell you. He will only tell you when you've done a lousy job, with feeling and volume.

I have never met anybody who feels 100 percent secure 100 percent of the time. Many of us at one time or another in our careers felt like imposters and that sooner or later we would be unmasked as not knowing what we were doing. Remember the first job you had? That first day at the staff meeting when everyone was talking technical and you were nodding your head like you understood, praying that someone wouldn't ask you a question because you hadn't a clue about what was going on? For many of us, these doubts quickly fade away.

Difficult bosses make us feel this way much of the time. These bosses tend to be egomaniacal, domineering, derisive toward subordinates, and loners who do it all themselves. I have found that many bosses who are egomaniacal generally feel that they are themselves incompetent. The only way they feel that they can look good is to make you look bad. Being domineering goes hand-in-glove with egomania. They are always in control, wanting to know all the details, and micro-managing your every decision, then complaining why you aren't done yet.

Do a 360-degree audit and get a disinterested third party to do the same. I was once railed at for my writing style and ability. After repeatedly getting beat about the head and neck about this, I sent a draft to three disinterested parties (a colleague, an editor, and an English teacher) and asked for honest evaluations. All three had praise. It put all of the negative comments from my boss in perspective. Once you complete the audit, build on your strengths and take definitive steps to improve your weaknesses. Then don't just file it away, revisit it every six months to track your progress. Believe it or not, this will build your faith in yourself. The bottom line here is faith.

Have hope in the future. Nothing lasts forever, good or bad. Persevere, hunker down, and just get through it. One day your boss will change areas of responsibility. You have to wait patiently. During this time you have to have hope. What you think about the future affects you today. Believe that the future can improve, and you will do things to bring it about. Conversely, if you believe that nothing improves, you will do nothing to change it.

For example, I was in a planning meeting and one person kept illustrating how everything we suggested would not work and fail. This person had no hope of the future. On the other hand, when Thomas Edison had failed for the 700th time to discover a filament that worked in the light bulb, a colleague commented, "We will never find a filament that works." Thomas Edison retorted, "I have not failed 700 times. I have not failed once. I have succeeded in proving that those 700 ways will not work. When I have eliminated the ways that will not work, I will find the way that will work." Thomas Edison had a positive view of the future even though the past and present were dim. The bottom line here is hope.

Trust in others. Many difficult bosses never develop a fundamental trust in others. As a result, they go it alone and do not depend on anyone else. In today's rapidly changing business world, this prehistoric approach does not work. The truth be known, it never really worked in the first place. After all, Batman had Robin, the Lone Ranger had Tonto, and Beavis has Butthead (just remember, I'm Beavis).

Today, most companies will agree that people are the company's greatest asset and resource. It is logical to then assume that trust in people is at the heart of good business. Those who try to do it all themselves generally complain that no one else pulls their weight. They look for evidence to confirm this belief. It isn't too hard. Do you ever catch yourself saying that if you want something done right, you have to what?—yes, you guessed it—do it yourself?

To improve your own trust quotient, select something to delegate, choose someone to delegate to, give them all the information they need to do a good job, and mentor them. You will be surprised what people can do if you give them the freedom to be creative. The bottom line here is trust.

Conclusions

Here are the three strategies—faith, hope, and trust. These three principles are at the heart of good personal and business relationships. In life and work, you do not have to make yourself out to be so big, because you are not that small in the first place.

Bouncing Back After Losing Your Job

Losing your job can leave you devastated, or it can lead you down a new path toward advancement and growth. The difference is in your perspective and your plan. Have you ever been fired or laid off from your job unexpectedly? If so, then you know what it's like to be hard at work one day and jobless the next. The main thing to remember is not to panic. Being out of work is common in today's fast-moving society. But thanks to the transient nature of the American workplace, you might be able to turn a setback into a step up on the career ladder. However, if you're going to bounce back after losing a job, you need a game plan—and the first step in that plan is to get organized. Don't consider yourself unemployed. Instead, employ yourself full time in your search for a new job.

Outline a realistic teemingly and don't delay. The faster you get started, the better and more confident you'll feel. Check the newspaper classified, but don't stop there. Most jobs are found through networking, and virtually everyone has the power to cast a huge net into today's well-stocked job pool.

Step by Step

Here are three steps that will prepare you to pound the pavement.

Self-Assessment

Take an honest look at your career, your accomplishments, failures, your likes and dislikes. Also think about your hopes and dreams for the future. Then put the past behind you and get excited about what's to come.

Changing jobs can mean a change for the better, so design your search to get the kind of position you really want. Begin by answering these questions:

- What are my strengths and weaknesses?
- What kinds of projects do I like most?
- What are my salary expectations and benefit requirements?
- Do I want to stay in the same industry?
- Am I willing to relocate?
- What lifestyle changes would I like?
- Are my skills up to date, and how can I improve them?

Materials and Research

It's more important than ever to package your work history in a compelling, powerful manner. Also, research companies before you interview with them and be knowledgeable about trends in their industry.

The Internet is a valuable research tool that can help you with these tasks, but it can be overwhelming. Some good starting points are:

- Careerpath.com (www.careerpath.com)
- Monster + Hotjobs (www.hotjobs.yahoo.com)
- Monster (www.monster.com)

These sites list employment opportunities and resources available on the Internet. Also check out individual company websites for targeted searches.

Don't forget the specialty sites, such as the American Society of Safety Engineers website (www.asse.org) and its Nexsteps SH&E Employment site (www.nexsteps.org), a great resource for SH&E professional looking for jobs.

You can upload your résumé into résumé banks at these major hubs. Just make sure your résumé is Internet savvy. You can do that by using software such as Resumail (www.resumail.com) and ResumeMaker (www.individualsoftware.com).

Strategy and Networking

Professional networking is probably the most effective, and most underused, strategy for finding a new job, so start attending those local chapter meetings and start talking to fellow professionals. And read. Study industry trade magazines for company news, trends, and employment ads.

Be sure to explore all networking avenues, including professional associations, executive recruiters, former employers and co-workers, and your list of professional contacts.

Keep an optimistic attitude. Remember, we live in an ever changing economy in which jobs appear and disappear on a regular basis. If you can manage financially, I suggest that you set your sights on finding the right job, not just an immediate paycheck.

Change Management

My father returned from World War II and took a job at Texaco in Port Arthur, Texas. Thirty-five years later he retired from the same company. He never changed jobs. That was then, this is now. Ironically, change is the one constant that exists in the business world today. It may even be the only constant.

In contrast to my father's experience, my first job was with Ford Aerospace. Ten years later they were bought by Loral. Later, I went to work for Union Texas Petroleum, two years later they were bought by ARCO, which was then bought by BP. Yet again, I went to work for the St. Paul Companies. Four years later they were bought by Travelers. I left Travelers and now work for an oil and gas company.

Consider how quickly things change. Freshman now entering college have:

- Always had a PIN number
- Never gotten excited over a telegram, a long-distance call, or a fax
- Always gotten black and white photocopies, not purple mimeographs
- Always known about HMOs

Twenty years ago there were 50,000 computers; today over 50,000 computers are installed daily. In 1995 there were 300 million emails per day; in 2008, there were 210 billion emails per day (http://www.techwatch.co.uk/2009/01/26/emails-reach-210-billion-per-day/). To further illustrate, there are 17 new web pages created every three seconds, and who knows how many blogs per day. There are more computer-literate first graders than there are computer-literate CEOs.

As a result, the need to adapt never ceases in today's business world. Typically, managers assume there is no change, leaders anticipate change, and followers resist change. No matter how seamlessly change is engineered, it isn't easy or natural for anyone. The key is to decide early what you are going to do. If you aren't going to adapt, leave as soon as possible. If you are going to adapt, eschew the old company colors and don the new. Have a "can-do" attitude and be the team player who adapts.

Some of the helpful practices of adaptability are:

- Let the past be the past. There are many good memories and successes. Let them remain there. Don't be the person who continually says, "When I was at company XYZ, we did it this way." You are no longer at company XYZ, you are at company ABC and they probably do things differently. Demonstrate your flexibility.Learn the difference between when to adapt and when to hold steady. If you adapt, does it move you off your mission? If so, hold steady. If you adapt, do you violate your core values? If so, hold steady. Is the desire to adapt not in the team's best interest? If so, hold steady. Otherwise, you will need to adapt.
- Maintain an open mind to all possibilities of change. During a time of change there are many things in question. Before you leave, wait and see what the change ends up being. It may even be better than before. If you don't wait and see you will never know. Be open-minded that the changes may be positive rather than assuming that the changes will be negative.
- Don't ride a dead horse–dismount. We often remain in the past harping on things out of our control. This is extremely unproductive. My comment is either get over it or leave. Don't sit here and make it worse for everyone else. Why keep riding a dead horse?

  Figure 4 lists a few thoughts on ways that companies insist on riding dead horses.

Figure 4. Dead horse strategies

- Change the bylaws so that the horse can't die.
- Rename the dead horse.
- Buy a stronger whip.
- Arrange a visit to other sites to see how they ride dead horses.
- Meet with the dead horse to discuss productivity.
- Harness other horses to the dead horse to increase speed.
- Conduct training on how to ride a dead horse.
- Have the CEO declare that no horse is too dead to ride.
- Appoint a committee to study the horse.
- Establish a quality circle to develop uses for dead horses.
- Promote the dead horse.
- Move the horse to a new location.
- Provide daily status reports on the dead horse.
- Terminate all live horses to redefine productivity.
- Hire a motivational speaker to inspire the dead horse.
- Hire a consultant to study dead horses.
- Hire another consultant to refute the consultant's conclusions.
- **The best thing to do when riding a dead horse is to dismount.**

With change so prevalent in today's business world, why are we so resistant to change? When you get accustomed to a work routine you get comfortable with a set of expectations. Then when change occurs it upsets all the previously-predictable events of the present and the future. Things that we wrestle with are:

- The fear of the unknown. We move from what we know to what we don't know. When a new business unit is forming and we are asked to go to a different office in town, a different state, perhaps even a different country, the fear of not knowing what to expect can be scary. The bottom line here is that we are people of habit and when our routines are disrupted fear often ensues.

- We've never done that before. When we do our jobs we get into routines. When we get a new job assignment we've never done before there is a question in our minds as to what the right answer really looks like. Is what we did really the right answer? Will I look the fool for doing it wrong? Until we get into a routine there is some trepidation.
- Leaning too much on past successes. When remembering the successes we accomplished earlier, sometimes we rely on them too heavily. If you just completed a big project last year, now you want rest on your laurels. This can be good as long as you don't obsess and rely only on past successes. Remember, yesterday ended last night. What this means is that, it is a good thing that we have performed in the past, but we must also be committed to performing in the future and continuing to prove yourself.
- Demoralization. When you get too attached to a routine, a way of doing things, a job, an employer, and change takes all this away, it can be demoralizing. Understanding that change occurs and that we need to look at things with a fresh perspective allows us to discover things we may not have been able to discover before.

On the other hand, there are those who embrace change. They love change. In fact, if change doesn't occur they institute their own change, sometimes by changing jobs. Why do people embrace change? Well, here are few things that I have learned:

- Try something new. Some people live for trying something they have never experienced before. While some have never left their geographical area, some live to go to different states or countries. In work they look for a new business unit, a new software program, a new process, a new whatever. If it is different, they want to try it.
- Look for an opportunity. Rather than looking at change as a bad thing, they look at change as an opportunity to learn, to grow, to excel, and to get better at what they do. They see

changes as opportunities to enhance and elevate rather than to deflate.

- Learn something new. Some people seek change to learn something new about themselves, the company, their career, or a new skill. They welcome and embrace rather than criticize it.
- Get out of a rut. Some people abhor the "same-old-same-old" rat race. They embrace change as a way to get out of the old rut and to revitalize their lives. Where some live for the rut, some live to break out of the rut.

In the end, we are either growing or stagnating. Which one are you? Are you still wearing the same kind of glasses you did twenty years ago? Have you gotten into a life-defying spiral so orderly that the slightest change annoys you to a migraine headache? Think about instituting some change. How about reinventing yourself every four years? Change your hair style, try different glasses, take a college course on something that interests you, just do something different. If you don't tweak yourself every so often, you won't grow

Success Tips

Know When to Leave a Company

1. You no longer believe in what you're doing or in your company's mission.
2. You're no longer growing and learning in your job.
3. You suspect that your job will be eliminated or relocated.

Plan Your Career

1. Never rest on your laurels, continually pursue new technologies and opportunities.
2. No one will plan your future for you.
3. Set a five-year plan, that is, write it down, and evaluate it every year as to where you are.

4. If possible, identify several paths to follow, for example, commercial versus government; insurance versus industry; academic versus consulting.
5. Diversify wherever possible.
6. Get as many related professional credentials that relate to your field.
7. Publish, publish, publish—but only when you have an aha, not when you have a so-what.
8. Present at local, regional, and national technical conferences.
9. Write about the ideas you have for books that no one else has talked about before.

Savvy Advice on Surviving a Corporate Merger

1. Develop a plan.
2. Maintain your professional integrity.
3. Be tactful with peers.
4. Watch what you say about the merger.
5. Assume personal control of your career path.
6. Manage your emotions.
7. Assume a positive leadership role.
8. Realize that some things never change.
9. Find new rules for success.
10. Accelerate your networking activities.

What Office Politics Are All About

In my twenty-plus years as an employee in the corporate world, I've been a consummate observer. I've seen just about everything there is to see. I've been on the upside and downside of mergers, acquisitions, joint ventures, and divestitures. As a young SH&E professional, I always wondered how people climbed the ladder so rapidly and stayed aloft so long. It wasn't just the star performers—in fact, few of the stars made it to positions of power and influence. I always hoped, as a Type A personality SH&E professional, that superior technical knowledge was the key to organizational success. It took me a while of spinning my wheels until I finally realized it wasn't just technical ability, but something quite intangible. The proverbial, "It isn't what you know, but who you know," began to find a place in my five-year plan. This went against the grain of everything I had ever learned about succeeding in business.

I saw many incompetents rise to the top and stay there, much to the chagrin of my pay-for-performance, Protestant, work ethic. It wasn't hard work after all, it was a combination of ability and who you knew. My goal was not to stay too long in middle management. In fact, I once posted in my office cubicle, "Temporary office of a rising young executive." I had my five-year plan, but no mentor to guide me through the maze of corporate politics. One wrong move and I'm gone or pigeonholed for life. Being someone who abhorred failure, I had to find a solution. As an engineer, I wanted to call it political engineering. But I knew that would be a trite oversimplification. So I called it "the politics of SH&E."

What is contained here is something I have not found taught in school, pontificated on in a continuing education class, or shared by any individual, even in SH&E. It is all through the school of hard knocks. Through repeated failures and trial and error, I have compiled material that no one wants to share. These are the strategies I have employed to successfully get things done and move up in the organization.

Ethics

Everything discussed from here out is ethical and legal. What we will do is toss out any conventional wisdom of how business operates as the only way to get ahead. What you will get is a roadmap on which you can fill in the blanks according to your situation to move up in the organization "using" politics rather than "fighting" politics. You will see how you can use the company's stilted, unsympathetic, unyielding, black hole infrastructure to move ahead rather than take the answer "no" and submissively move on, as the song by Tennessee Ernie Ford says, "Another day older and deeper in debt."

Today's conventional wisdom dictates that we avoid offending people in the organization at all costs. Rather, I advocate acting aggressively and effectively in a way that everyone benefits from. As a result, there is some risk. I know that most SH&E professionals are risk-averse, so this may go against the grain and you may find it difficult to step up and actually do some of these activities.

Ethics are not to be taken lightly. This is one of those things that can end a career. Just ask the people at Enron and other companies who failed to take heed of ethical principles. Ethics come from many directions, including Safety, Health, Environmental, Human Resources, and Legal. These issues must be dealt with carefully, confidentially, and professionally. Further, some of the things we see can can potentially end others' careers when it comes to non-compliance, particularly in SH&E. ASSE's Code of Conduct is shown in Figure 5.

Figure 5. ASSE Code of Professional Conduct

Code of Professional Conduct

Membership in the American Society of Safety Engineers evokes a duty to serve and protect people, property and the environment. This duty is to be exercised with integrity, honor and dignity. Members are accountable for following the Code of Professional Conduct.

Fundamental Principles

1. Protect people, property and the environment through the application of state-of-the-art knowledge.
2. Serve the public, employees, employers, clients and the Society with fidelity, honesty and impartiality.
3. Achieve and maintain competency in the practice of the profession.
4. Avoid conflicts of interest and compromise of professional conduct.
5. Maintain confidentiality of privileged information.

Fundamental Canons

In the fulfillment of my duties as a safety professional and as a member of the Society, I shall:

1. Inform the public, employers, employees, clients and appropriate authorities when professional judgment indicates that there is an unacceptable level of risk.
2. Improve knowledge and skills through training, education and networking.
3. Perform professional services only in the area of competence.
4. Issue public statements in a truthful manner, and only within the parameters of authority granted.
5. Serve as an agent and trustee, avoiding any appearance of conflict of interest.
6. Assure equal opportunity to all.

Approved by House of Delegates June 9, 2002

The Board of Certified Safety Professionals (BCSP) Code of Conduct is sown in Figure 6.

Figure 6. BCSP Code of Ethics and Professional Conduct

CODE OF ETHICS AND PROFESSIONAL CONDUCT

This code sets forth the code of ethics and professional standards to be observed by holders of documents of certification conferred by the Board of Certified Safety Professionals. Certificants shall, in their professional safety activities, sustain and advance the integrity, honor, and prestige of the safety profession by adherence to these standards.

Standards

1. Hold paramount the safety and health of people, the protection of the environment and protection of property in the performance of professional duties and exercise their obligation to advise employers, clients, employees, the public, and appropriate authorities of danger and unacceptable risks to people, the environment, or property.

2. Be honest, fair, and impartial; act with responsibility and integrity. Adhere to high standards of ethical conduct with balanced care for the interests of the public, employers, clients, employees, colleagues and the profession. Avoid all conduct or practice that is likely to discredit the profession or deceive the public.

3. Issue public statements only in an objective and truthful manner and only when founded upon knowledge of the facts and competence in the subject matter.

4. Undertake assignments only when qualified by education or experience in the specific technical fields involved. Accept responsibility for their continued professional development by acquiring and maintaining competence through continuing education, experience and professional training.

Figure 6. BCSP Code of Ethics and Professional Conduct (Cont'd)

5. Avoid deceptive acts that falsify or misrepresent their academic or professional qualifications. Not misrepresent or exaggerate their degree of responsibility in or for the subject matter of prior assignments. Presentations incident to the solicitation of employment shall not misrepresent pertinent facts concerning employers, employees, associates, or past accomplishments with the intent and purpose of enhancing their qualifications and their work.
6. Conduct their professional relations by the highest standards of integrity and avoid compromise of their professional judgment by conflicts of interest.
7. Act in a manner free of bias with regard to religion, ethnicity, gender, age, national origin, sexual orientation, or disability.
8. Seek opportunities to be of constructive service in civic affairs and work for the advancement of the safety, health and well-being of their community and their profession by sharing their knowledge and skills.

Approved by the BCSP Board of Directors, October 2002.

Your Roadmap

You probably didn't think I meant this literally, but I really do mean a roadmap. I got this idea from one of my early bosses, who told me about the difference between the formal organization chart and the informal organization chart. As I charted this out, I added functions that had potential political indicators. What you see here is what evolved from this initial concept. Before you can apply the techniques explained here, you'll need an organized, clear picture of the lines of power and influence that exist within your organization. Start with a blank sheet of paper. When done, you'll have a directory of all the people and their interrelationships.

Start with the top-ranked people in the company. Write their names in a row across the top of the page. Make a circle or square about two inches across around each name, allowing space for comments (Figure 7).

Figure 7. Step 1 in your roadmap

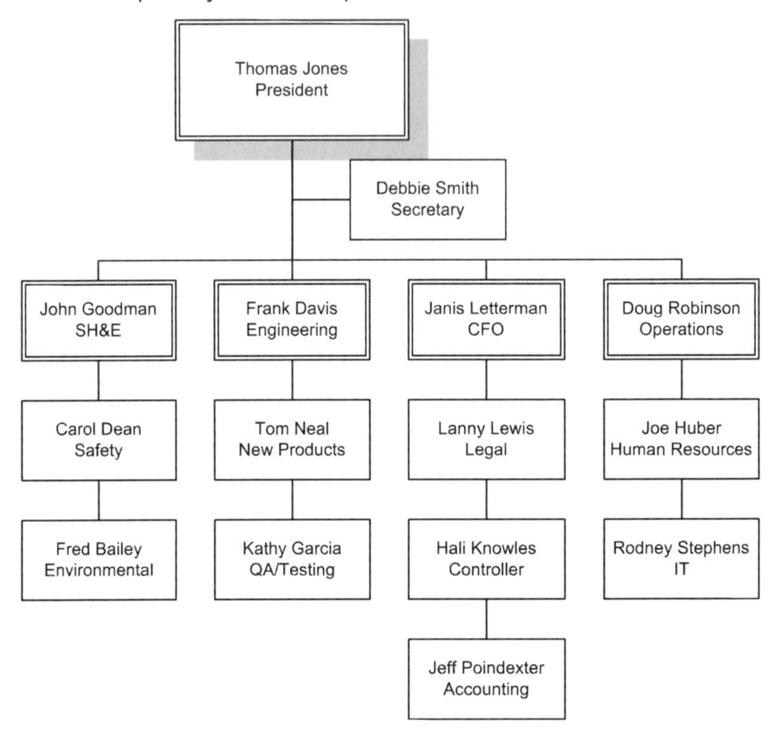

Beneath the top-ranked people, write the names of the people who report to them, and continue down the chart past your level so that there is enough room for one or two ranks of names beneath you. If you have to list more than 25 people, it will be easier to write each name on a separate slip of paper, arrange the slips, then copy the names onto the master sheet when you have arranged them according to rank.

Your next step is to write one or two of your impressions and comments about each person in his or her square (Figure 8). Be sure to include information about the type of relationship you have with each one, interests you share, and potential areas of conflict.

Figure 8. Step 2 in your roadmap–traits of people

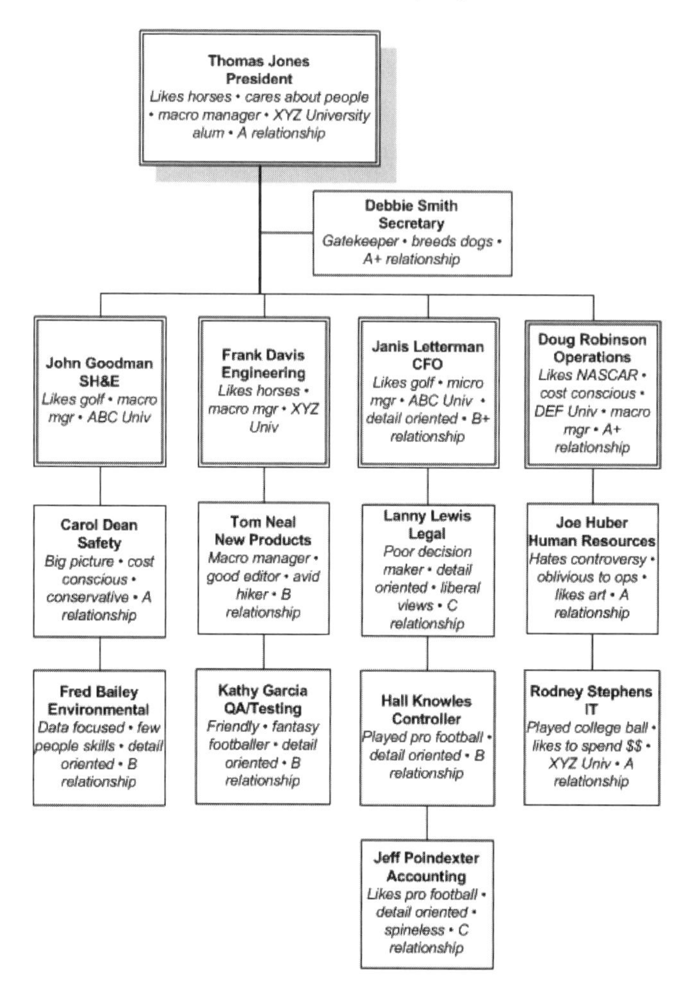

Next, and this is the crux of the matter, indicate all reporting relationships by drawing lines between the circles using a blue pen (some of the lines will have to be curved). Using a red pen, draw lines to indicate any personal or strong political alliances you have perceived between people, which may or may not coincide with reporting lines. See Figure 3 for an example of strong political alliances.

Figure 9. Step 3 in your roadmap–strong political alliances

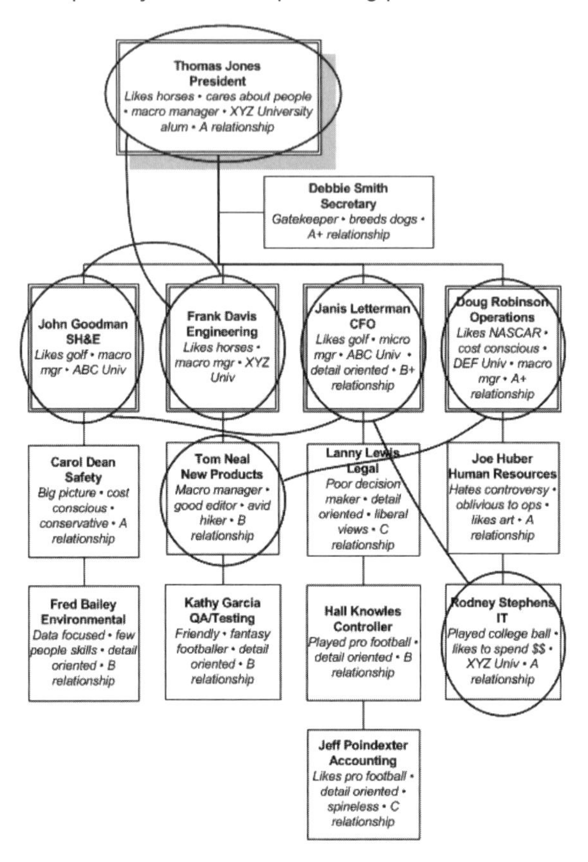

Next, on each line, write one or two facts about the reporting relationship (Figure 8). Use a pencil so you can update information when necessary. For example, John reports to Mary, but only on budget; Sue is too completely dependent on Elaine, and can't make a decision without consulting her.

When it is all put together it can get a little messy. That is why I showed you in steps how it should look, depending on the organization, traits, relationships, and alliances. Figure 9 shows what it might look like with all three on one chart.

Don't leave this roadmap lying around, it is a valuable political tool. I recommend keeping it at home. I would call it a war room chart of sorts. It allows you to visualize instantly all the relationships and cross-relationships that exist in your workplace, putting you at an instant advantage over people who try to carry around all this information in their heads.

It gives you a view of the areas where you can exert influence on people, often without talking to them directly. It can even help you chart a path to your next promotion or analyze power factions within your support staff to accomplish your five-year plan.

Don't fail to update your roadmap frequently. When you're dealing with a large number of individuals, you'll be surprised how frequently new information emerges. This is especially true after major changes in the company such as mergers, acquisitions, divestitures and joint ventures.

You'll be amazed at the power edge this tool will give you.

Your Personal Power Inventory

To solidify your position and win advancement in today's competitive business climate, you have to apply all the abilities and assets you have at your disposal. This is a journey, not a destination. That is, it is not stagnant but dynamic, ever changing. You have to be on your toes, poised for change at any moment or you will lose your advantage. Worse yet, you could lose your position or that promotion you've been waiting for. Be careful not to fall into the following traps.

Working harder. Political advancement does require that you do an excellent job to earn respect. But throwing hard work alone at the problem is not enough. This is definitely against conventional wisdom, especially for SH&E professionals.

Relying on a limited support base. You could build a system made up of a limited group of friends from your college, community, or elsewhere. But using this approach exclusively is too limiting and cuts off more support than it provides. I graduated from a university known for its alumni networking. I use this network whenever possible, but it is not my sole source of networking.

Trading favors. The old back scratching can win self-advancement. This approach has its usefulness as an isolated technique, but relying on it exclusively turns you into a caricature of a slimy snake oil salesman.

These techniques may be useful, but should not be relied upon solely to move up. There are other more useful techniques.

Your Keys to Power

When thinking politically today, you have to uncover the many tools you possess that you 're not using, overcome any resistance you may have to activating them, then apply them aggressively in a structured method to get ahead.

Figure 10. Your personal power inventory

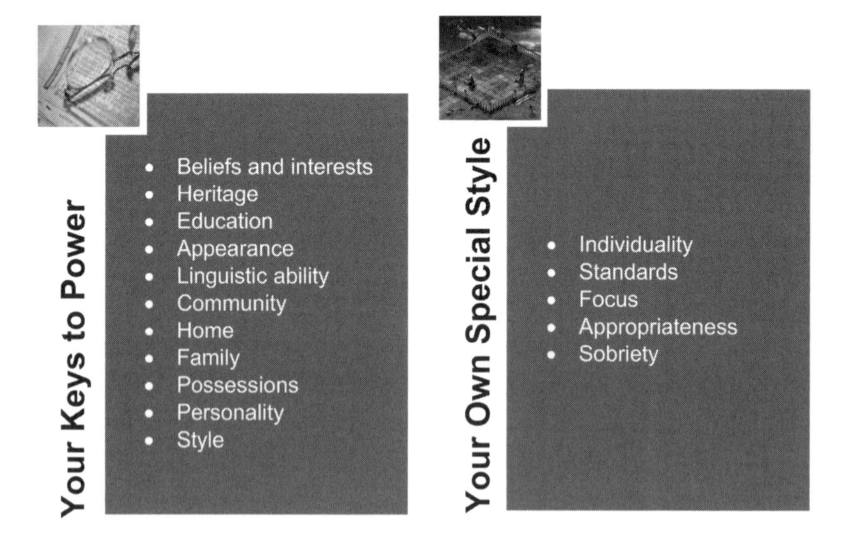

Consider the assets shown in Figure 10 that you may not be using:

Beliefs and interests. If you have strong political, moral, spiritual, or other beliefs, don't keep them a secret on the job. Hobbies, sports, political and religious beliefs, and other interests can also gain you an immediate set of connections.

Heritage. While discriminating *against* people makes you an anachronistic legal liability, don't be reluctant to take advantage of your cultural, religious, racial, or national background when selecting a firm to work for or when building alliances. For example, I work for a company in Minnesota, where a large Norwegian contingent settled many years ago. As you might imagine, with a last name like Hansen—which is like Smith anywhere else—I can fit in pretty easily (you betcha!). Also, I grew up in Southeast Texas near the Louisiana border. I have leveraged the Texas heritage (Howdy!), the East Texas heritage (Hi, y'all!), and the South Louisiana heritage (Cómo se va?).

Education and schooling. Look for well-placed alumni of your alma mater. Review what you studied to be sure you are not neglecting skills or interests that could help you rise on the job.

Appearance. An average appearance shouldn't slow you down in business. You can't rely on good looks exclusively, but count them as an asset that can help you look the part for a leadership role.

Linguistic ability. If you grew up speaking a second language or acquired proficiency through study, you have a valuable tool. Use it to build ties to others who speak the same language, and look for opportunities to help your business move into foreign markets or establish ties abroad.

Community. Do you live in the same community as colleagues or members of upper management? Sharing a commute or some talk about town politics and other community concerns can strengthen a valuable alliance. However, a warning here: sharing too much or the wrong information about yourself can jeopardize your future.

Home. If you can use your home to host power parties, why not?

Family. Don't be reluctant to call upon your relatives for advice or help. And sharing conversations with colleagues about children's schooling and other family matters can establish a comfortable common ground with fellow workers. If your spouse can offer advice to your colleagues on various concerns such as real estate, investments, and so on, don't be shy about taking advantage of it.

Possessions. Do you collect paintings, cars, or stamps? Don't hesitate to make these interests known at work. You may find a network of other people who share your interests.

Personality. You *can* build an alliance based on this elusive factor. Do you have a keen sense of humor? Are you a hard-boiled skeptic or grump? Believe it or not, both of these traits can work for you. Look for people who share your outlook and stop by to laugh about some recent event, or exchange grumpy thoughts. Most of us in SH&E made our marks by livening up SH&E meetings with a bit of humor. Use it to your advantage.

Style. Are you a sophisticate, a jock, a chic dresser, a tweedy rumpled type, or something else? You may have to make some stylistic modifications to survive in certain businesses, but look for colleagues who share your approach. You'll experience a higher level of comfort and better communication with these people.

Your Own Special Style

In order to advance, you'll need to be taken seriously. What is the style that draws leaders upward toward the boardroom, and how can *you* acquire it?

Individuality. Don't try to anticipate what other people might like, simply espouse enthusiastically what you *do.* While sharing pursuits with others is important, having an involved, enthusiastic life is what will make you attractive to others, not whether they share all your interests and pursuits. In an effort to cultivate corporate style, many people try to emulate the attitudes, interests, and outlooks of all their colleagues. They try to fit in, and finally do just that, right into pigeonhole-sized niches they built for themselves. I once worked for a company where the president of the division was a two-fisted drinker and a cigar smoker. All of his direct reports took up drinking and cigar smoking. He retired and his replacement was a bicycle enthusiast. The running joke was that it would be quite a challenge for his direct reports to switch from drinking and cigars to biking. Trying to conform to corporate culture is a mistake, if your company is inhospitable to your style and way of thinking, trying to conform will only make you a fish out of water. You're in the wrong company.

Standards. Carefully define what your business and personal principles really are, then stick by them staunchly, no matter how difficult the circumstances. If you are fully committed to the profitability of your department and it's time to let some workers go, fire those who are least capable, regardless of your personal feelings. Ask yourself, "Do you want to make friends or earn *respect?*"

Focus. The ability to concentrate completely on a task and get it done is central to gaining advancement. And unlike intelligence, this ability can be developed. Focus is also the most important part of what is often called "executive presence." To see it in action, observe the person of power in a meeting. While others are veering off toward other issues, this person will not be dissuaded from his or her goals.

Appropriateness. Treat serious issues with a serious attitude. This is something I struggle with regularly. I enjoy work and want to have fun there. After all, I am spending eight to ten hours a day there. However, while telling a few jokes to blow off steam may break the tension of difficult decisions, it will make you appear uncommitted.

Sobriety. Nothing will undermine your credibility more quickly than having a few drinks at lunch and returning to the workplace smelling of alcohol. Further, drinking and acting out of character at a holiday party or picnic is *verboten*. What you do at home is your own business, but what you do at work *is* business. Even if colleagues and superiors drink at company dinners and such, hold the urge and "just say no." It will take you miles. By the same token, inappropriate romantic or sexual liaisons will jettison you right off the corporate ladder.

Political Mistakes from Which You Will Never Recover

An unfortunate byproduct of political structures is that they will network *against* you as well as *for* you if word gets out that you've made a blunder. Be especially careful of the following political gaffes:

Bad timing. Before asking someone for help, consider whether your relationship is ready to bear the weight of what you are requesting. If you haven't progressed beyond establishing that you and your contact share an interest in travel, it's too soon to ask for help in winning a new opportunity or promotion.

Stepping on toes. Before you look to people for support, consider the positions they occupy. If a new opportunity or set of responsibilities is up for grabs, asking the head of another department for help in winning them may be a serious misplay if he or she is considering making a play for them also. There is an exception: if your alliance with them is exceptionally strong and well maintained, asking for support can result in their dropping out of the competition, sometimes without resentment. Alternatively, it may open up discussion of a mutually beneficial approach to the problem.

Relying on foundationless alliances. If you're friendly and cordial with another manager, that's not enough to enable you to rely on him or her for strong support. Be sure that there is a variety of other supporting factors and a history of cooperation to add a foundation to your relationship.

Overusing an alliance. The more valuable an alliance, the *less* often you should use it for help with big projects. A strong ally will be working silently for your interests in many small ways without your knowing it every day. If you walk into a highly-placed colleague's office too often to ask for a big favor, the quality of your connection will erode quickly. Think of an alliance as a rechargeable battery. After using it, you have to reenergize it before tapping into it again. After getting a favor, return a favor in a big way or do some high-level stroking to keep the alliance running strong.

Posturing. When trying to win support or influence someone, it is *very dangerous* to infer that your project should be supported because you are strongly connected to someone at the top. First, the inference is a threat. Second, if the top person turns out to be a less-than-staunch supporter in this instance, you're going to lose face with everyone. Third, everyone will know that you are strongly connected to the executive anyway, and will weigh that information in deciding whether or not to support you. Blowing your own horn only makes you a blowhard.

Letting alliances lapse. If the very person who can help you is someone with whom you used to enjoy close political ties, but you've been neglecting him or her lately, it's a serious blunder to count on picking up your former close relationship where it left off. It is also a serious misplay to start currying favor suddenly by taking the person out to lunch, calling him or her up to chat, etc., before making a pitch for support. The best thing to do is maintain close ties to your colleagues, letting none of them lapse. The next best thing to do is come clean. Tell the person that you are aware you haven't been in close touch lately, but that you have good memories of the projects you've shared in the past, and that you are calling to ask for support. Strengthen this approach with an immediate offer of support for one of your contact's current concerns.

The Influence Game

Fighting Intimidation: How to Spot Deceitful People

It's great to solidify your power and build your political network. The downside is that you are going to become the target of carping from envious and often malicious people who will try to intimidate you, bruise your self-image, and make you feel your growing power is of no significance.

Your best strategy, the only one really if you're going to win, is to study power plays with objectivity and learn as much as you can. The wonderful truth is that they offer you the chance to learn a tremendous amount about your enemies and opponents, far more than you could if they were silently dreaming up plots against you. They're making the mistakes, and you're reaping the benefits. Here are the most common types of detractors and what you can do about them.

Self-aggrandizers. These opponents routinely exaggerate accomplishments or political clout. When the company president says good morning to them in the hall, they come by to tell you they just had a policy meeting. Over a period of time, look objectively for discrepancies between what a person says and what you see is happening. Perennial self-aggrandizers are generally not taken seriously by anyone. Unless there are some mitigating factors (the posturer is the president's nephew, for example), you need not take them too seriously, either.

Insinuators. They engage in Iago-like intrigue. *Example:* Your opponent stops by to tell you that your boss had to rewrite your last report because it was disorganized. A *real* snake may even infer that he or she is on your side and offer to help you complete projects to avoid further trouble. Your defense is to confirm any reports directly with all those concerned. Insinuators *are* dangerous because they'll take their intrigue on to other people after they have failed with you. They are also often highly magnetic people who gain credibility with other people for a while. Your best defense is to wait until this type of person plays the same card too often and is exposed for what he or she is. *Mistake:* Warning another employee that the insinuator is a chronic liar can drag you into a morass of charges and countercharges.

Sleight-of-hand artists. This type tries to throw you off by providing some spurious information. For example, just before a meeting, he or she will tell you: "That idea of yours will never get approved. The vice-president proposed the same thing last year and it got shot down." It's hard not to be fooled once or twice by this type, especially if he or she is clever. Your defense is to make sure you're not fooled from then on.

Strategists. When you're cooperating on a project with this type, he or she says there's no pressure and starts off at a sluggish pace. When you follow the lead and start slowly, he or she secretly catapults forward, finishes far ahead of you, then turns in his or her work early to make you look like a snail. Your defense is to do your work at your own pace and not to be intimidated. Just because the strategist thinks he or she is earning points with upper management by these antics doesn't mean it is the case.

Bullies. People who try to bulldoze or intimidate you are hard to analyze. Your best defense is to assess their clout objectively and firmly stand your ground despite all attempts to sway you.

How to Double Your Chances of Getting Others to See Things Your Way

The more you want something you're about to negotiate for, the more you're likely to weaken your chances by over-preparing and over-documenting your case. You're far more likely to win if you make *smart* preparations rather than *hard* preparations. Review the following checklist:

Keep things simple. Distill your ideas into a few sentences and keep supporting paperwork minimal for the negotiating session itself. Don't be the one that drones on, communicating in five minutes what could be communicated in one. Don't be the one who causes people to say, "Is there a point somewhere in our near future?"

Structure your presentation. Reveal your biggest idea immediately, without preambles. Don't say, "I have a proposal so far-reaching I predict it will save the company $2 million this year alone." Just present your idea clearly and let predictions follow. If your first points are well received, go on to make your second and third points, but resist the temptation to keep throwing in lesser and lesser points, they only weaken your stance.

Anticipate objections. Be sure to do your homework on how management is likely to react to your proposal. Research how it has treated similar proposals in the past, and anticipate objections your opponents are likely to make.

Build in some sacrifices. To protect the meat of your proposal—the things you really don't want to see cut out—add a few chips you're willing to give up. You can then bargain them away and end up with the parts that mean the most to you.

Don't try to score all the points. Your position will actually be strengthened and cooperation much more likely if you accede to some ideas from people who are critiquing your presentation.

Meet resistance flexibly. When you care a great deal about a project, you may tend to overreact when anybody raises questions. Treat objections with humor and try to foster an atmosphere of cooperation; it implies that your project is already underway. If negotiations stall completely, try to ascertain what aspect of your proposal is *really* bothering the other side. Don't become argumentative, but try to read between the lines of what is being said. As a last resort, consider throwing one of your sacrifice chips away to get things moving.

Don't give away the store. Accept a certain number of modifications to your plan or idea, but don't let it get converted into another person's project that you're now expected to supervise. If you see this happening, offer to withdraw the entire proposal for further development.

Rise above politics. It's funny advice to be giving here, but try not to consider whom your opponents are when you're actually making a presentation even if they are right there in the room. The more important your proposal, the more vital it is to appear to have the interests of the entire company at heart, especially when top management is in attendance. Don't betray any aggressive feelings toward your opponents or act condescendingly toward them.

Typically, there are three kinds of people in your organization:, those who support you, those who don't support you, and those who are on the fence. You don't need to spend too much time convincing those who support you or those who don't support you. The ones who support you are already allies. The ones who don't support you don't care what the issue is; they will oppose any proposal. The ones you need to focus on and convince are the ones on the fence. This will sway your proposal to the majority.

Transitioning into SH&E from Other Closely Allied Fields

As stated throughout this book, in order to transition into SH&E, make a plan. Map it out and execute it carefully. Review it regularly and modify it as necessary. Basic elements of the plan if you are trying to transition within your company include:

- Taking classes from colleges or universities, whether local or by distance learning.
- Attend conferences on your own time and expense if necessary.
- Attain a certification like an OHST, ASP, or a CSP.
- Meet with those in the company who are in the SH&E department. Find out what they are doing and how they do it. Pick up the lingo and research the materials.
- If possible, find a mentor in SH&E. Meet regularly to discuss progress and plans.
- Take opportunities at brown bag lunches to see how things are going.
- This sounds basic but, go to the safety meetings and participate. Research the materials afterward.
- Volunteer to do a safety meeting outside your realm of expertise.
- Let management know you are interested in an SH&E position. One of the first things management does is look inside the company walls for some interested and trainable.
- Keep your resume updated with all of your extracurricular SH&E learning activities.
- Publish an article closely allied to your field where it overlaps with SH&E.
- If a merger or acquisition occurs, let the other company know of your interests. It could open the door for an internal transition.

Here are some examples of opportunities for transitioning from different fields to SH&E.

Current Field	Transition to:
Health sciences and nursing	• Ergonomics • Wellness • Worker's Compensation cases with health-related issues • Chemical exposure, monitoring and understanding MSDSs • Radiological exposure monitoring and testing • Hearing testing and conservation • Noise measurement • Program development
Quality	• ISO/ANSI Management standards • Management Systems development and implementation • Auditing • Metrics
Human resources	• Employee complaints regarding ergonomics • Wellness • Worker's Compensation cases with health-related issues • Program development • Drug and alcohol testing
Operations	• Safety committees • Job safety analyses • Understanding the culture • Program development • Drug and alcohol testing

Current Field	Transition to:
Security	• Workplace violence • Homeland security issues • Job safety analyses • Program development • Drug and alcohol testing
Risk management	• Claim management • Insurance strategies • Worker's Compensation, automobile, property, pollution, general liability, etc. • Program development • Contracting and subcontracting • Legal aspect of SH&E
Clerical	• Statistical data and metrics for SH&E using spreadsheets and presentation software. • Website/Intranet maintenance and update. • Following up with incident reports. • Updating and understanding the SH&E databases

Working closely and expressing an interest in SH&E opens the door to potentially moving into SH&E. Coupled with the items listed earlier regarding the basic elements increases the chance of making the move, either within the company or with another company.

Survival Skills

How to Reprimand a Subordinate

The classic rules of reprimanding a subordinate have always been to do it behind closed doors and to avoid personal attack. They're good rules. Bend them only when the employee is a repeat offender you're trying to embarrass or motivate to quit, or when there are other special circumstances.

Preparation is the key to effective reprimands. Arm yourself with appropriate documentation for the points you wish to make. The key is to document everything. You may choose to use a small bound notebook to take notes in. Date the observations and concerns. It is always wise to get Human Resources involved. Otherwise, you'll have to stop in midstream and resume later when you have your information together. That's awkward and it gives your subordinate time to think up a defense.

Presenting Your Case

Here's how to capitalize on your position of strength:

Wait until you're under control. While you may gain an edge by issuing a reprimand while you're angry, it can also work against you because you're more likely to forget to make key points or think clearly about solutions to the problem. If you want to appear angry, act angry but wait until you are really calm internally and in control.

Be specific about the offense. Make your complaints **one** at a time. Don't try to soften any blows. Be straightforward.

Give the person a chance to explain. There may be extenuating circumstances, but it's more likely that the employee will simply own up to the problem and start suggesting ways to improve.

Don't harbor a grudge. Unless the employee is a habitual offender, say that the book is closed on the issue, provided that the situation changes. If it does not, the fact that you've treated the employee fairly and seen no improvement only serves to add to your clout the next time you issue a reprimand or move to have the offender terminated.

How to Handle the Press Like a Pro

Frankly, I couldn't tell you how to attract positive publicity for your firm. You'd better talk to a public relations expert about that. But what I *can* tell you about is how to handle a hostile press when you're under fire for something your firm has or has not done.

First, do yourself the favor of claiming some basic privileges for yourself: the right to be treated politely and fairly, the right to have a chance to say what you want to, and the right to remain in control of the situation, including the right to turn your back and walk out of the room if your other rights are not being respected.

Coping with Emergencies

With luck, you'll have time to assess the situation and gather background information before you're on the spot. However, in an executive's nightmare, there may come a time when you're suddenly before the cameras trying to explain why your company's oil tank caught fire, or why your firm is closing a plant that employed a large number of people.

The *worst* thing you can do is lie to the press about what happened. Just as harmful is to issue an opinion based on skimpy information in the hope that you'll later be proven right by the facts.

The press certainly won't want to hear the standard line, "That matter is currently under investigation." However, in reality, it is often the safest and most honest thing you can say. But honesty will only get you so far with reporters, who may try to trip you up with the following ploys:

The "What if...?" question. Example: "What if the fire spreads and fatalities result? Will your firm make restitution to the families of victims?" Don't ever let yourself get pulled into speculation. Reporters can trap you in a seemingly unimportant one, and then lead you into more damaging statements.

The "yes or no" question. Example: "Did your company take adequate SH&E precautions, yes or no?" This is an overt ploy to make you look bad, and it will work. To limit the damage, say: "We will be issuing a report containing that information shortly," but don't pick some arbitrary date when an answer will be forthcoming.

The "number one priority" question. Example: "What is your company's number one priority in fighting pollution?" If you say you're reducing your plant's output of airborne pollutants, the press might criticize you for ignoring water purity restrictions. Defense: Say, "We are attacking many major concerns, including ..."

The "off the record" question. Always respond to a reporter as though your statements will become a matter of public record. Odds are that they will.

The "either/or" question. Example: "Either your company wants to keep the citizens of this town employed, or it wants to close down the factory. Which is it?" This is an attempt to paint you into a corner. Your best defense is to point out that the "either/or" case is spurious. Time permitting, make an explanation of issues that confront your company at the time.

The "multiple choice" question. Example: "Will you replace outmoded equipment to meet new EPA guidelines, repair the old equipment, or just shut down temporarily?" Simply because the reporter is supplying the options doesn't mean that you can't ignore them and supply some of your own.

The "statement" question. Example: "You obviously don't want to spend money on community development projects." You can defend yourself by converting the statement into a question: "If you're asking about our current programs, let me explain them for you..."

The "second guess" question. Example: "How do you think your competitors will respond to your new product line?" Defend yourself by saying, "You'd better ask *them* that question."

The "policy statement" question. This can occur when you go before the press armed with information on a specific issue, and then some reporter asks: "What is your company's stand on the environment?" To avoid a lengthy silence or an ill-conceived reply, tie your answer into the matter you came prepared to discuss, and point to it as an example of your company's outlook. Questions like that present a golden opportunity to display your background research, and tell of past activities that show your firm in a positive way.

How to Report Bad News Positively

How do you give bad news, such as discussing a wage freeze or layoffs, and still have people believe in your personal worth and good intentions regarding SH&E?

To boost your odds for success:

Avoid surprises. Keep people informed ahead of time that these problems are a major management concern.

Avoid mixed messages. Give the bad news and don't try to soften it. Telling people that they're being let go but that the company will now turn a profit will not earn you any points with anyone.

Give full information. While you want to keep your central theme clear and understandable, don't fail to offer as much backup information as you can.

Encourage free communication. Saying, "My door is always open," after you let ten people go may make *you* feel better but probably won't do the same for anyone else. Better yet, schedule individual appointments with any concerned employees and take care to address their personal concerns and needs. Don't just recite the same speech to each of them.

Don't overstate your sorrow. Even though you *may* feel as badly as the people who are getting the bad news, telling them that you're as miserable as they are will never ring true.

The Pros and Cons of Hiring a Friend

If you have decided that hiring a friend is always either a very good or a very bad idea, you haven't given the matter enough thought.

The bottom line is that by hiring someone, whether a friend or a stranger, you are establishing a *professional* relationship. When considering a friend, consider the same factors you would when making any hiring decision but add the pluses and minuses that your friendship might bring to the professional relationship.

Consider your friend objectively. Are you sure that he or she is *really* the best person for the job? How about other candidates? If your friend is not the most knowledgeable or experienced for the job, do strength of character or the advantages of your relationship outweigh any shortcomings?

Consider the nature of your friendship. Look at areas in which you communicate well, but also consider areas of friction. It's reasonable to expect your professional relationship to mirror these tendencies. The bottom line is this, preventing problems *before* you hire someone can avoid both professional and personal friction.

Build in an escape chute. Establish a mechanism for terminating the business relationship if it is not working out, and for regularly sharing views of how the liaison is going. If possible, it may be desirable to bring your friend in on a temporary basis for a predetermined period of weeks or months, and then cement the relationship if it is working out well.

Consider the benefits and detriments of having a friend on the job.

Knowing that you can count on someone can be a tremendous plus. However, it can also cause friction with other staff members, who will resent the fact that a colleague has a close personal relationship with you.

Weigh the political implications. Like it or not, you'll be operating as a team at least in the public view. Consider the ramifications in the light of the political relationships you've already established. Is hiring your friend worth the potential damage it may cause?

Personal Patterns of Success

Decorating Your Office So You Look Like a Leader

Your office should function well for *you.* It should look clean and efficient, without piles of paper or other clutter. But you must also pay close attention to your office's atmosphere. Does it convey an air of seriousness and efficiency? Does it look like a place where some valuable thinking is taking place?

I have visited the offices of many important people in my day, and they all shared the qualities I mention above. Curiously, very few of them achieved the same results in the same way.

To decorate your office as a leader would, obey the following rules:

Express your personality. Your office should reflect your interests and pursuits. Include things that appeal to you. Looking for items to please other people only waters down the personality you're trying to convey, and can make you seem wishy-washy. Never be ashamed of your tastes. If you love your classic 1967 Corvette, put a picture of it on the wall, even if your colleagues all have reproductions of Impressionist paintings.

Include things you understand. If you put up a painting or a framed quote you like and know about, you'll create a positive impression when you explain it to people. If you put up something you're supposed to like but really don't, you'll seem pretentious. **Use** help wisely. As you climb the ladder, someone will offer to decorate your office. If you can keep your interests at the center of the plan, fine. Otherwise, refuse and keep your office the way *you* like it.

Avoid the commonplace. Don't decorate your space with two or three innocuous framed posters that the mailroom brings you—it's another way to ensure anonymity. By the same token, avoid cute cartoons at all costs. Displaying a cartoon of a monkey saying: "I Hate Mondays!" will only make people view you as a monkey who hates to work.

What a Well-Positioned Mentor Can and Cannot Do for You

A mentoring relationship is actually a close political alliance in which the flow of information goes more strongly in one direction than in the other. This usually occurs when a more senior person recognizes some special potential in a subordinate and singles him or her out for cultivation.

My own belief is that a broad network of cultivated relationships is likely to be more beneficial to a younger manager than a strong political alliance with one senior person would be. However, if there is one powerful person who seems to take a special interest in you and wants to further your ideas, or who is very well placed and knowledgeable in your field, the advantages of cultivating a relationship may outweigh the lack of political mobility you'll have to accept.

You don't have to be crazy about somebody to want to establish a close alliance with and learn from him or her. Sometimes respect and an attitude of interest are all that are needed to form the backbone of a mentor-student relationship. But in most successful cases, the relationship has been nurtured and strengthened by political cultivation. A well-placed mentor will speed your progress and, with luck, teach you some important things. However, you should be aware of the following pitfalls:

Picking the wrong mentor. In your eagerness to establish ties with a bigwig, don't establish ties with somebody who is not well connected, who doesn't know as much as you expected, or who is about to leave. Do your homework carefully before establishing ties. You're better off being unallied than tied to the wrong person or camp.

Getting hamstrung. When you have a mentor, you can worry too much about taking risks out of fear that your actions will reflect badly on your ally. So you become an ineffectual clone who waits for suggestions of what to do. *Defense:* Try to find a mentor who encourages independent effort, and continue to act with autonomy in most of your projects.

Getting reined in. You have an idea you think is terrific, and propose it at a meeting. To your surprise, your mentor is furious because he or she wasn't consulted, or thinks it is a bad idea. *Problem:* When you agree to a mentoring relationship, you are also agreeing to a certain amount of control.

Political limitations. Having a mentor not only cuts you off from having close ties to your mentor's opponents, it also creates disturbances with colleagues at your own level. How would you like to have a colleague who shares your rank and title, but who has the ear of top management while you do not? Of course, your colleagues are free to advance themselves politically, but you must calculate the risk of schisms among your coworkers.

Ugly surprises. Just because your mentor tells you all about certain things doesn't mean he or she will tell you that you're getting bypassed for a promotion, or that a key assignment is going to somebody else. The biggest surprise of all is when your mentor leaves. Never forget that you are a subordinate, and there are limits on the information you'll receive.

Blurring of roles. Always remember that your mentor is your superior. In almost all cases, it is unwise to confide that you are unhappy with your job, or that you are seeking a new one. Also try to avoid gossip, and continue to act with autonomy in your projects.

In matters of joke-telling, sharing personal information, and social interaction, always let your mentor take the lead in establishing guidelines. For example, never be the first to suggest that you have lunch.

Burnout. A close relationship can eventually degrade if you become a threat or a mental burden to the person who has taken you on. Be willing to distance yourself when necessary and don't overtax the relationship by asking for advice too often, or stopping by when you have nothing or something minor to discuss.

How to Make Points with a Boss Without Obviously Being on the Make

Your relationship with your boss is likely to be taken as a microcosm of your future with the company. If you manage it well, it is taken as a barometer of your potential in the firm. Here are some vital pointers for managing this all-important relationship effectively:

Question your viewpoint. If you think your boss is either beyond reproach or incapable of doing anything right, you have failed to make an accurate or useful assessment of his or her abilities. And you are not meeting his or her needs. Go back to square one and take an objective inventory of what he or she can do.

Learn how to format information. As a subordinate, one of your major duties will be to supply your boss with facts. Don't format the process the way *you* like it. Try to meet your boss's preferences. If he or she prefers a daily morning chat to a stack of memos, oblige that need.

Understand his or her personality and your own. For the relationship to work effectively, you may have to subjugate your own style. To accomplish this, you will have to do some self-analyzing. If you're fun-loving and your boss is glum, superimposing your style on the relationship can result in disaster.

Learn when to make proposals and when to keep quiet. Especially in the early stages of a relationship, it is vital that you offer only your best and most fully-developed ideas. Spattering your boss with a buckshot hail of suggestions in the hope that some will hit the mark brands you as someone who is right only two or three percent of the time.

Anticipate pressures. Pay attention to how and when your boss feels pressure most acutely. *Examples:* Is it when he or she is facing a meeting with a particular high-ranking person? Is it when he or she prepares reports? Try to throw your support at these areas first. Don't blindly believe that your priorities match those of your boss.

Defend your boss. Never gossip about your boss's shortcomings; it's far too dangerous a game. If your boss has fallen on hard times with upper management, then you will have to make an appraisal of his or her situation and act accordingly. In all but the most extreme cases, defending your boss is your own best defense.

Test Your Promotability: Are You *Really* a Logical Candidate for a Move Upward?

Just because you're doing an efficient job of handling most of the facets of your career, don't assume that you're promotable. When you get passed over for a big promotion, you'll end up saying: "How could they do that when I motivated my staff so well?" or, "How could they do it when I just got my first invitation to the boss's executive barbecue?" To make it to the top, you've got to keep building up the areas where you're weakest.

Don't say I didn't warn you. Take the following test. Any areas in which you answer *no* can scuttle your chances, so start working on them now.

- Have I established a strong individual presence and profile?
- Are my speech and communication skills as strong as I can make them?
- Have I consistently supported colleagues, taken an interest in their projects, and helped them achieve results?
- Have I demonstrated a strong ability to confront and solve problems, or have I just skirted them?
- Have I demonstrated an ability to perform under pressure?
- Have I consistently completed management's priority assignments before my own?
- Have I completed projects that demonstrated my specific strengths and abilities?
- Have I defined my long-term objectives, and do I understand the role my current activities play in attaining them?
- Have I supported my superiors by cheerfully taking on work in difficult periods and offering good advice?
- Have I achieved a high degree of political clout that reinforces the strengths outlined above and seals the promotion?

Nine Strategies for Successfully Playing Organizational Politics

When you deal with superiors, peers, and subordinates, they all see who you really are. Whether you are dealing with finance, facilities, janitors, shipping clerks, secretaries, receptionists, or scientists, there are nine strategies that make organizational politics quite simple. Here they are.

1. Be nice to everyone.

Don't believe the "nice guys finish last" stuff. In organizational politics, nice guys build supportive relationships with other people. Bulldozers and sharks make enemies, and enemies make your life miserable by resisting and sabotaging you. Be sincerely nice to everyone, not just the people you think can help you. People resent phoniness. You just never know when that janitor you speak nicely to every day will let you in your office that one day you forgot your key or that young professional who finds the fast track who suddenly becomes your boss.

2. Be a team player.

A team player is someone who helps the team achieve its goals, and helps other people achieve their goals. Be a star by making other people look good, rather than taking credit for other people's achievements.

3. Don't whine and complain.

Develop a reputation for being a problem solver. Anybody can complain about a problem. The really valuable employees are those who prevent or solve problems. Although the squeaky wheel may get the grease, coworkers resent a whiny employee.

4. Be visible.

You can't win office politics by hiding. You can't be an office rat. You must be involved and others need to perceive you as a valuable contributor at work. Get involved in solving important and highly visible problems at work. In the age of downsizing, many employees are shocked to learn that they lost their jobs because upper management didn't know what they were contributing to their organization. You not only have to do good work; others must give you credit for your good work and perceive you as a good worker.

5. Help your boss succeed.

This is part of being a team player. Your job should be to get your boss promoted. It is also a smart strategy because your boss is a major player in your promotability and in how upper management perceives you and your work. If you have a positive relationship with your boss, your boss is going to be more likely to support your career and help you advance.

Now some of you are reeling in disgust because you hate your boss. Well, get over it. You will have a more difficult time winning organizational politics if you openly declare war on or antagonize your boss. Remember, they probably have kept their job and power base because they know a few things about playing organizational politics.

You don't have to suck up to your boss, but you must nurture your relationship with him or her. If you disagree with your boss, do it privately. Be very careful, though, about embarrassing your boss in public or in front of your boss's superiors or staff. You don't want your boss trying to get even with you.

6. Be loyal.

As I have stated throughout this book, loyalty is a lost art. It is quite simple to be loyal. Primarily, avoid backbiting and backstabbing. Coworkers will support you if they believe that you will be there to support them. To get loyalty, you need to show your loyalty.

7. Be good at what you do.

Develop your expertise and competence. Show up on time and work hard for your full shift. To survive in the workplace, you also need to do be a performer. If others perceive you as a slacker or a poor performer, they will not support you. They will also resent your getting promoted before them.

8. Mind your manners.

Be polite and courteous. Avoid being sarcastic or putting other people down. When in doubt, err on the side of being gracious.

9. Make other people look good.

We already mentioned this concerning being a team player, but it bears repeating. People will support you when they believe that you make them look good. They will resent you if they believe you take credit for the work they do. Give credit to others. Sincerely compliment others. Help people look important and successful in front of the people who are important to them.

Sixteen Ways to Damage and Destroy Your Working Relationships

On the flip side, there are sixteen ways to tear down working relationships with coworkers. They are as follows:

1. Make insulting comments about their private lives.

Just when the relationship is being built up and trust is surfacing, slam-dunk them with insults. This is a sure-fire way to destroy a relationship.

2. Unfairly blame them for a negative outcome.

If you have ever seen the TV program, *The Apprentice*, when the going gets tough, the contestants blame their coworkers for the failures and bask in the glow of the team successes, taking all the credit. The lesson to learn here is when you don't cover their backs, they won't cover yours.

3. Make decisions without their input.

We live in a team world. Very few accomplishments are the result of individual performance. When you act autocratically, without the input from the team, you are setting yourself up for failure.

4. Interrupt them while they are speaking.

Boy, do I hate interrupters. This is the ultimate act of disrespect—that what I have to say is more important than what you have to say.

5. Look down on them or treat them negatively.

Second to interrupting is being condescending. Treating coworkers with disdain and disrespect goes miles toward destroying work relationships.

6. Spread false rumors about them.

Remember the old saying: "all rumors are true, and all rumors are false." The best bet is to not start or spread rumors at all. If someone wants to know about a coworker, let them hear confidential information from the source, not you.

7. Judge their work unjustly or unfairly.

Setting performance standards as a moving target is like shooting at a target in a dark room with no feedback, and will destroy working relationship in moments. Slanted, biased reviews of coworkers' products are a way to take leaps to destroying coworker relationships.

8. Criticize them unfairly.

Hypercritical coworkers are the stuff Tums and Rolaids are made of for the recipients of this mean-spirited act.

9. Insult them directly.

It is one thing to insult people behind their backs, but insulting them to their faces adds insult to injury.

10. Ignore them.

When they call, don't answer, when they email, don't respond, when they instant message or text you, ignore them.

11. Question their abilities or judgments.

Once you've appropriately delegated a task, look over their shoulder and micro-manage the task. Question every decision, no matter how minute. Give them the micro-manager blues and let them know you are in charge. Too bad you won't have time to do your own job.

12. Embarrass them in front of others.

Don't adhere to the adage "praise in public, punish in private." When you make a spectacle of them in front of the whole group you make your point. They won't argue with you again. Unfortunately, they won't likely open their mouths to contribute either, allowing you to fail in the future.

13. Talk badly about them behind their backs.

Backbiting is one sure-fire way to destroy relationships, especially if they find out from the office chatter.

14. Belittle their ideas.

Whenever they open their mouths with an idea, destroy it as a bad one since you didn't think of it.

15. Get other people to turn against them.

Share their personal idiosyncrasies with their coworkers so that everybody knows their faults. That will make them feel safe and secure. Yeah, right.

16. Keep them from expressing themselves.

Stop them midstream from communicating their ideas. Don't even let them finish their sentences. That is yet another way to destroy relationships.

Seven Habits of a Successful Politician

1. Practice good interpersonal skills.

Listen, pay attention, and encourage communication. We have two ears and one mouth for a reason.

2. Don't let anyone get something on you.

The best way to not let this happen is to not do anything worth getting. Acting with integrity and character are great ways to prevent someone getting something on you.

3. Be a producer, not a politician.

Paradoxically, not being a politician is the key. If you do your job and do it well, performers like this get promoted and protected.

4. Do favors for others.

This sounds oversimplified, but if someone needs a door opened, open it. If someone smiles at you, smile back. If you see someone's printout on the printer, deliver it on the way back to your desk. This isn't rocket science. If you want to be liked, you have to be likable.

5. Clearly define your job and role.

Understand we all live with the "other duties as assigned" clause in our job descriptions. But develop your roles and responsibilities and stick to it for the long view. Over time you will find that in between all the daily crises you will have accomplished all those things you wanted to do.

6. Expose dirty tricks used by others.

Don't be a tattletale, but make sure that the fact that others aren't playing fairly is brought out in the open.

7. Make other people look good.

Your job is to make others working with you look good, especially your boss. When you employees fail protect them, when they succeed give them credit.

Summary

Office politics is a dangerous game if you are not prepared, protected, and properly networked. I have presented somewhat opposing views, but at a minimum you need to do your homework and listen more than talk. When you talk more than listen, people learn all about you, but you learn little or nothing about them. Listen intently, as though you would be reciting it back to the individual speaking, take copious notes, and read between the lines. When you speak, do so concisely and with a strong point. Use the things that are a true representation of who you are to your advantage. You need to know who the power brokers are and the ones who are not. What you see here is over twenty-five years of observation and a few choice actions that have bolstered my career. Perhaps this will help you in moving your own career forward.

Success Tips

Understanding Office Politics

- Every company has them. Get to know how things happen.

Ethics

- Everything must remain above-board.

Your Political Roadmap

- Identify and map the real power structure of the management organizational chart.

Take a Personal Power Inventory

Identify your key attributes:

1. Beliefs and interests
2. Heritage
3. Education and schooling
4. Appearance
5. Linguistic ability
6. Community
7. Home
8. Family
9. Possessions
10. Personality
11. Style

Your Own Personal Style

Things that make you unique:

1. Individuality
2. Standards
3. Focus

4. Appropriateness
5. Sobriety

Political Mistake from Which You Will Never Recover

1. Bad timing
2. Stepping on toes
3. Relying on foundationless alliances
4. Overusing an alliance
5. Posturing
6. Letting alliances lapse

Fighting Intimidation

How to spot deceitful people:

1. Self-aggrandizers: They routinely exaggerate accomplishments.
2. Insinuators: The inference leader.
3. Sleigh-of-hand artists: Spurious information providers.
4. Strategists: Work at your own pace.
5. Bullies: Stand up for yourself.

How to Get Others To See Things Your Way

1. Keep things simple.
2. Structure your presentation. Start off with the big bang of your proposal.
3. Anticipate objections. Be ready with plausible solutions.
4. Build in sacrifices. If they know you are willing to give, so will they.
5. Don't try to score all the points. A win is a win. You don't need a shutout.
6. Meet resistance flexibly. Don't overreact.

7. Don't give away the store. Cede little by little, not all at once.
8. Rise above politics. Don't hold a grudge. It is just business.

Survival Skills

1. How to reprimand a subordinate:
   - Behind closed doors
2. Presenting your case:
   - Wait until you are under control
   - Be specific: have examples documented and ready
   - Give the person a chance to explain
   - Don't hold a grudge
3. How to report bad news positively:
   - Avoid surprises
   - Avoid mixed messages
   - Give full information
   - Encourage free communication
4. The pros and cons of hiring a friend:
   - Stay objective
   - Consider the nature of your friendship
   - Build in an escape chute
   - Consider the ups and downs of having a friend on the job
   - Weigh the political implications

Personal Patterns for Success

1. How to decorate your office so that you look like a leader:
   - Express your personality
   - Include things you understand
   - Avoid the commonplace

How to Make Points with the Boss Without Obviously Being on the Make

1. Question your viewpoint.
2. Learn how to format information.
3. Understand his or her personality and your own.
4. Learn when to make proposals and when to keep quiet.
5. Anticipate pressures. Pay attention to how and when your boss feels pressure most acutely.
6. Defend your boss. Never gossip about your boss's shortcomings; it's too dangerous.

Testing Your Promotability

How do you answer these questions?

1. Have I established a strong individual presence and profile?
2. Are my speech and communication skills as strong as I can make them?
3. Have I consistently supported colleagues, taken an interest in their projects, and helped them achieve results?
4. Have I demonstrated a strong ability to confront and solve problems, or have I just skirted them?
5. Have I demonstrated an ability to perform under pressure?
6. Have I consistently completed management's priority assignments before my own?
7. Have I completed projects that demonstrated my specific strengths and abilities?
8. Have I defined my long-term objectives, and do I understand the role my current activities play in attaining them?
9. Have I supported my superiors by cheerfully taking on work in difficult periods and offering good advice?
10. Have I achieved a high degree of political clout that reinforces the strengths outlined above and seals the promotion?

Nine Strategies for Successfully Playing Organizational Politics

1. Be nice to everyone
2. Be a team player
3. Don't whine and complain
4. Be visible
5. Help your boss succeed
6. Be loyal
7. Be good at what you do
8. Mind your manners
9. Make other people look good

16 Ways to Damage and Destroy Your Working Relationships

1. Make insulting comments about their private lives.
2. Unfairly blame them for a negative outcome.
3. Make decisions without their input. We live in a team world.
4. Interrupt them while they are speaking.
5. Look down on them or treat them negatively.
6. Spread false rumors about them.
7. Judge their work unjustly or unfairly.
8. Criticize them unfairly.
9. Insult them directly.
10. Ignore them.
11. Question their abilities or judgments.
12. Embarrass them in front of others.
13. Talk badly about them behind their backs.
14. Belittle their ideas.
15. Get other people to turn against them.
16. Keep them from expressing themselves.

Seven Habits of a Successful Politician

1. Practice good interpersonal skills.
2. Don't let anyone "get" something on you.
3. Be a producer, not a politician.
4. Do favors for others.
5. Clearly define your job and role.
6. Expose dirty tricks used by others.
7. Make other people look good.

Being a Business Person Who Understands SH&E

Far too often I run into SH&E professionals who simply lack the big picture of how we fit into the business. I know our jobs are altruistic, but if management doesn't see us as business people, we will always be labeled as the "SH&E geek" and nothing more. That is why I suggest writing a mission and vision statement with support objectives that link into the company's mission, vision, and objectives. If we can demonstrate how we bring the two together, we become an added value to the business team. What follows are ways to demonstrate your business value to the company and lessons I have learned over the years.

It's not Personal, It's just Business: Business Lessons for the SH&E Professional

Seeing things not for what they are but for what they might be creates opportunities. It's easy to take a great performer and keep it performing. It is quite another thing to take a poor performer and turn it into a good or great performer. Taking a poor performer and turning it around can make a statement for your career. Get the reputation of a turnaround artist and you will never go begging during your career. Get the reputation for making lemonade out of lemons and people will want to follow you your whole career.

Never be afraid to state your true worth. When someone asks what you are worth, be able to state with confidence, and based on performance, what your true worth really is. Many are shy in stating their worth, and as a result they never really demonstrate their value to the organization. It may sound arrogant, but if you couch it in terms of value to the organization people will see what you bring to the bargaining table.

You're the one in charge of your learning curve. When you are given a new assignment or you take a new job, no one is going to learn the new system but you. If you dive in and learn all the issues in the new job or assignment, you demonstrate the quick study you are and that you have become integrated into the team.

When it comes to assessing people, don't always believe what you hear: form your own opinion. When you hear what others have to say about the new addition to your team, your new boss, or the new CFO, don't rely on it. Oftentimes people don't get along because of mere personality conflicts. Since your personality is different, perhaps you won't have the same problems as others do. Give it a try, you may be surprised at how well you get along with the new person that you were told was one massacre short of Attila the Hun.

The key to mastering a new job is finding a mentor. Whether it is a professional mentor or a company mentor, a mentor can help you learn the ropes of how things get done in the most efficient manner. A mentor can help you learn in the informal network of getting things done and help you avoid the land mines and find the pathways around the hurdles that people put in place.

Don't mistake asking questions for ignorance; asking questions often shows that you know what you're doing. Many take questions as a personal affront to their abilities, when in reality the questions are being asked to get an understanding of how business is being conducted. Don't take it personally, they are just trying to learn your method of conducting your job. They just may be able to help you do it better.

The key to a good presentation is to keep from getting caught up in the bells and whistles and cut to the chase. We, as SH&E professionals, tend to overload decision makers with data because we don't know how to couch our arguments succinctly. Try to develop a memo on three-quarters of a page rather than five pages. Avoid the 30-page document if possible. If you can't shorten the data, craft a one-page executive summary. When you can, develop a graph that is easy to understand that communicates 30 pages of data in a compelling form to clearly and succinctly explain your argument. Remember, details are for backup only.

You need a strong sense of self image to manage a weak boss. A weak boss will manage you to failure, and as a result, you will take the heat for organizational weaknesses. A weak boss may also attack you for failures that he or she should have dealt with, but didn't because he or she didn't have the business acumen to convince peer management. That is yet another reason to pick your bosses carefully.

Get an up close and personal reading on your new boss. That means listening more than talking, asking other subordinates, asking people outside the company, but most importantly, from your own interaction. Identify your boss's strengths and weaknesses, demeanor, and when is the best time to bring good news and bad news to them. Watch, listen, take notes, and learn how to best interact with your boss. It could mean the difference between success and failure.

Have clear goals that your boss understands. We all need to have business goals, but when our goals are different than the ones we have been given from our boss, trouble ensues. Make sure your goals are clear—that is, written down. That is a good reminder, every day, when times are difficult and we are conflicted between competing goals. Make sure you are on the same page as your boss.

Always anticipate your next move. Keep an eye on the organization and think about the possibilities if things go well and if they don't. Be prepared when opportunity prevails and be standing in the wings when things fall apart. Keep a confidential plan and review it frequently and modify it when necessary.

Be prepared to defend your turf. When reorganizations occur, rightsizing, downsizing, or whatever, when part of your turf gets allocated elsewhere, that's a bad sign. Defend all your disciplines. If you don't, your half-life as a manager just expired.

Have a base level of good performance to fall back on. We all have bad days, life happens. But when you do, it is less painful when you have performed well to date, rather than snuck in just below the radar.

There is no room for emotion in business. There's no crying in baseball or business. No matter how emotional you get in meetings or confrontations, keep your emotions to yourself.

In the absence of trust, even the most generous act seems questionable. This is a two-way street. It can happen because of you or from the people you deal with. When people have been mean and difficult to deal with, and then are suddenly nice to you, suspicions begin to mount as to motive.

Never use your title as the sole answer to a question. I have seen all too often, when asked a particularly probing question, the reply was, "I'm the SH&E director!" My reply is, "So what? Just answer the question." This is usually a defensive response to sidestep the question, hoping you will be intimidated and move on. Don't fall for it.

If your boss is a bully, you're probably stronger than he or she is. If your boss is mean and nasty and manages by volume, then just understand that this too shall pass and try to minimize the damage to your career.

It is appropriate to go over your boss's head when he or she is managing to fail. However, you had better focus on the impact to the company and not make it personal or emotional. Get all the details down and present your case in a strong, succinct, clear, and unemotional case.

When someone hands you an opportunity, don't be afraid to take it. SH&E professionals often avoid risks in business because as professionals, our goal is to minimize the risk at work. It is often difficult to switch gears from not taking risks to taking risks. It is difficult to learn that taking business risk to improve the workplace is a good thing. Don't be afraid to take an opportunity when it presents itself, it may just further your career.

Never forget that tomorrow's janitor could be tomorrow's manager. I have seen where companies require young professionals to rotate between business units so that one day they can manage the whole company. If you treat them poorly, trust me, they will remember it.

You might have the title, salary, car, and perks, but you don't have the job until you've proven it. Regardless of where you are in the organization, where you've been promoted to, you still have to prove yourself the old-fashioned way. You have to earn it. That means paying your dues at each level.

Take the time to get to know every one of your employees. Your employees will go to the ends of the earth for you if you get to know them as people first and employees second. When an employee goes into the hospital, do you make sure all the benefits are taken care of so the spouse isn't worrying about paycheck issues, or are you worried about replacing the employee? Your organization knows where you fall on this issue and indicates the kind of organization you run.

Once you're the boss, you don't need to know all the answers, just where to find them. How often do people come into your office with problems expecting you to fix them? What do you do? How about turning the question around and asking, "What do you think you need to do?" Make them think it through. Make them come up with the answer. They don't need to check their brain at the door. Make them contribute to the solution.

Showing authority can be as simple as doing what needs to be done. When you have a bad employee who needs to be disciplined or fired, do it. That will build the morale of the organization. They will see that you want to protect them from bad apples. There's an old saying, "when no one gets along with Bob, perhaps Bob is the problem." Deal with it in a timely manner or your organization will suffer, sometime dramatically.

If you make a mistake, don't make excuses and don't be afraid to ask for help. We all make mistakes, but the ones who don't own up to it lose the luster to their reputation. Owning up to your mistakes is hard, blaming others is easy, but is far more damaging in the long run.

A bad apple with a bad attitude can compromise the team. When you have a cancer in your organization you must cut it out quickly, or it will infect the whole group and possibly destroy the cohesiveness of the group.

Winning isn't everything; *how* you win counts just as much. Winning and taking the high road is a great path, winning and taking the "I told you so" low road can create more enemies than goodwill.

Good Employee Rules

If one wins, the whole team wins. When one person shines, the whole team shines. We are all in this together. Celebrate individual successes that build the team up.

Everyone has to appreciate everyone else's job. One person may be the industrial hygienist, another the emergency response coordinator, and a third the SH&E professional. All three have different roles equally important to the organization, and all must be equally appreciated. This builds camaraderie and esprit de corps. This is what builds momentum. Keep it up and there is nothing you can't accomplish.

Be in the know. Know what's going on in the office, in the field office, and in your discipline, including emerging technology and regulatory trends.

You've got to want to be there. This is the proverbial "fire in the belly" passion. When you lose the passion, quit and find something you do enjoy—or retire. Otherwise you will be a drag on your team.

Bring a solution, not a problem. All too often employees bring only problems to be solved by management. Make a habit of bringing a problem wrapped in at least one or two candidate solutions.

Make a decision, and stand by it. Some people would rather die than make a decision, and when they do, they vacillate between the options, waiting for a majority vote. Make a decision and stick to it. If it needs to be modified down the road, make the adjustments and keep moving.

Become an expert in your field. Keep learning. I've heard people say, "I know everything about ..." Learn everything you possibly can about your field. Be the expert.

Your Commitment to Profits

As SH&E professionals we should be committed to earning money for the company. For us this means minimizing losses as well as how we spend the money the company has allocated to us in our budgets. Every dollar counts. We can either use it wisely or spend it foolishly. The choice is ours. Our superiors will come to understand quickly just how we view the company's money. Do we treat it as if it is our own or spend it freely? Is it a valued resource or something we need to spend before we lose it?

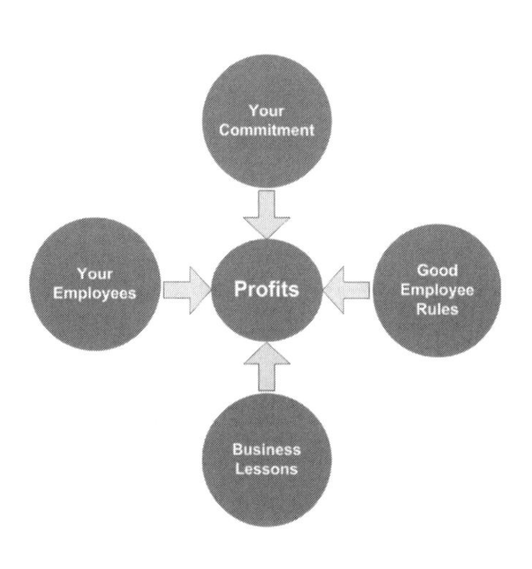

What does "the best" mean? Are we satisfied with the status quo or are we willing to be the best and make our company the best at what they do? Are we just putting in our time or are we relentlessly pursuing our vision? Answering these questions through our actions tells the whole story to our peers and superiors regarding the kind of businessperson we are.

Never apologize for focusing on profits. What? We are SH&E professionals, how can we focus on profits? Every dime we spend in our budget is a cost. Every dime that we give back to the company goes to company profits. For example, does everybody really need a top-of-the-line color printer whose life cycle cartridge costs far exceed the printer costs? Why not have a network printer for every group of five to ten employees? After all, how many color copies do we really need on a daily basis? If you want to share with other departments, that is a possibility as well, as long as they pay as they go. When you negotiate items in your budget, work the vendors. If you are buying bunker gear, negotiate the best possible deal, getting vendors to compete. The rationale is simple: you want them to make a profit, you just don't want them retiring on your order.

Strategic versus non-strategic costs. Anything that contributes to the bottom line is a strategic cost. Strategic costs are those that improve the bottom line, such as conducting internal Phase I Environmental Site Assessments before buying or selling properties, implementing automated or semi-automated systems that improve safety and productivity. Any other costs are non-strategic costs. Our goal is to eliminate non-strategic costs such as rent, consultants, state-of-the-art computer-related equipment, and office supplies.

Don't over-quantify things. We as SH&E professionals tend to over-quantify everything down to the gnat's eyelash when a more general quantification will do. Don't over-engineer your costs. In business, we rarely know the exact cost because sometimes it is a moving target. Be satisfied with understanding the costs that you can identify. It may take too much time to understand the exact cost in a business where minutes count. Further, don't quantify things you already know or that add no value.

Don't over- or under-delegate. This is a delicate balance. Good managers do one percent of the work but add 50 percent of the value to the organization. Weak managers study their business in detail and manage their business in general. Good managers study their business and manage the important details.

Strategic versus nonstrategic time. Any activity that contributes to your goals and profits is strategic time. These activities are task-oriented, customer-driven, and improve the business with which the company is involved. Nonstrategic time is everything else. Our goal is to focus on the future.

Every cost is up for grabs. Every year that you use the same contractor or vendor without going to bid, you risk experiencing a phenomenon that can be called "corporate creep." Every year a vendor increases rates and fees based on the cost of doing business and inflation. The result is that you pay more for the same product or service. Scrutinize those invoices and compare them to last year's and make sure you aren't getting caught in this negative effect. Better yet, go to bid frequently to prevent the corporate creep from occurring.

Cut costs first, ask questions later. When you think your budget is being eaten up by a black hole, cut things you don't need and cut deeper than you think you need to cut. Many of the things you think they will gripe about missing won't surface. The things they really want, they will come to you and ask for it and justify the reasons they need it.

Employees

As an SH&E manager, if you never fire employees, you can't have an excellent business. There are those who would prefer to coast until retirement, merely warming a seat and not pushing the envelope as a SH&E professional. Those who are marginal and poor performers need to be informed of the expectations of the job. Many believe employment itself is an entitlement. When you fire them for performance, the other performers on the team get fired up, knowing that you expect and demand their best. For the non-performers, you either fire them up or they update their resume while waiting for the other shoe to drop or put their resume on the street. Either way you are pushing excellence.

Setting salaries. Some SH&E managers are in the position to set salaries. Those who do should consider that pay should be connected to performance, not seniority. For employees with greater impact on the bottom line, their salaries reflect that value. If you see a superior value, recognize it and reward it. When hiring an SH&E professional, a particular applicant was working in a position that was less than his value. When we made an offer, we based it on his worth, rather than on his current salary. He interviewed with a competitor who made an offer based on his previous salary. Obviously, he accepted our offer. This was a great way to start off, with implied respect and loyalty, before the first day of work.

Never give regular bonuses. Second to the expected "retiring-in-place" job salary entitlements, are expected regular bonuses. Employees come to expect bonuses if you give them every year, rather than based on performance. If they don't get them every year, they know it is because of performance. Don't get caught in the trap of regular bonuses. It is hard to wean employees off the entitlement expectation.

Motivating employees. In the end, you should provide a creative and forward-looking work environment. One of the best ways to ensure motivation is that employees can stick with you, work hard, learn a lot, have fun, and be well-paid.

Run lean and mean. When hiring employees, make sure you are at least six months behind the manpower curve. It takes a lot of time and effort to hire people. It is truly painful. However, even more painful is laying off employees when the business wanes and there is no other choice. Make sure the demand is there before you hire.

Close the outside contractor loop. Consider performing tasks in-house rather than contracting them out. One sure-fire way to save on budget is to make your staff do the tasks they are competent at doing rather than paying millions of dollars for contractors to do them. It gets your employees closer to the day-to-day business and saves money at the same time.

Summary

Well, there you have it. I've included some methods to focus on the business and look for opportunities to leverage SH&E, implement SH&E while being cost-conscious, from contractors to employees. Perhaps these principles have opened your eyes to learn more business principles. I suggest reading business books, the *Wall Street Journal*, *Investor's Business Daily*, the business section of your local newspaper, and books about financial statements. Just reading the front pages of free online newspapers is a great start in beginning to expand your business horizons.

Success Tips

Business Tips

1. See things not for what they are but for what they might be creates opportunities.
2. Never be afraid to state your true worth.
3. When it comes to assessing people, don't always believe what you hear, form your own opinion.
4. Don't mistake asking questions for ignorance; asking questions often shows that you know what you're doing.
5. The key to a good presentation is to keep from getting caught up in the bells and whistles and cut to the chase.
6. You need a strong sense of self image to manage a weak boss.
7. Get an up-close and personal reading on your new boss.
8. Have clear goals that your boss understands.
9. Always anticipate your next move.
10. Be prepared to defend your turf.
11. Have a base level of good performance to fall back on.
12. There is no room for emotion in business.
13. In the absence of trust, even the most generous act seems questionable.
14. Never use your title as the sole answer to a question.
15. If your boss is a bully, you're probably stronger than he/she is.
16. When someone hands you an opportunity, don't be afraid to take it.

17. Never forget that tomorrow's janitor could be tomorrow's manager.
18. You might have the title, salary, car, perks, etc., but you don't have the job until you've proven it.
19. Take the time to get to know every one of your employees.
20. Once you're the boss, you don't need to know all the answers, just where to find them.
21. Showing authority can be as simple as doing what needs to be done.
22. If you make a mistake don't make excuses and don't be afraid to ask for help.
23. A bad apple with a bad attitude can compromise the team.
24. Winning isn't everything; *how* you win counts just as much

Good Employee Rules

1. If one wins, the whole team wins.
2. Everyone has to appreciate everyone else's job.
3. Be in the know.
4. You've got to want to be there.
5. Bring a solution, not a problem.
6. Make a decision, and stand by it.
7. Become an expert in your field. Keep learning.

Your Commitment to Profits

1. As SH&E professionals we should be committed to making money for the company.
2. Never apologize for focusing on profits.
3. Don't over-quantify things.
4. Don't over or under delegate.
5. Every cost is up for grabs.

6. Cut costs first, ask questions later.

Employees

1. If you never fire employees, you can't have an excellent business.
2. Pay should be connected to performance not seniority.
3. Never give regular bonuses.
4. Be creative in motivating employees.
5. Run lean and mean.
6. Close the outside contractor loop.

How to Find Out What You're Worth

Knowledge is power. That's true in everything from politics to gambling, and it's also true in compensation issues. But few safety professionals know their market value, despite a wealth of available salary information. People may be shocked to know that, in spite of downsizing over the past five years, good salaries can still be found.

Employers have an interest in keeping their workers in the dark, especially with the job market fluctuating from hot to cold as often as it does. If you're getting annual salary increases of four to five percent and you're in any reasonably warm field, you're probably doing okay. However, not keeping pace with the new people at your company is one reason companies are having a field day raiding talent from rivals.

Learn What You're Worth

So where to start? Major business publications, including *The Wall Street Journal*, publish annual pay surveys, as do many major professional groups, including the American Society of Safety Engineers and the American Industrial Hygiene Association. Check out the trade magazines, such as *Industrial Safety & Hygiene News*, *Occupational Hazards*, and *Occupational Safety and Health*. The Bureau of Labor Statistics also provides useful information on job titles and salaries. When you attend conferences, check out the positions from entry to seasoned and look at the requirements for each position and the salaries stipulated. Look for the education, experience, and credentials required and map them to salary. Also, check out the Nexsteps section on the ASSE website for the same information. Build a matrix of the details and be prepared to explain it to potential employers.

But don't stop there. Look through general-interest publications, such as *Money*, *Business Week*, and *Success*, that profile successful executives. The articles often include useful compensation information.

The Internet also contains considerable data. A good place to start, experts say, is JobStar (jobstar.org), which offers dozens of online salary surveys.

Job search networks such as ExecuNet (www.execunet.com), and NETSHARE, Inc. (www.netshare.com), provide salary guidance to their subscribers. These networks list hundreds of managerial job openings (including safety), most of which include salary information.

But the surveys are just a starting point. You must also factor in industry, company size, skills, geography, and of course, demand. Most of the sky-high salaries reported for MBA holders, for instance, come from consulting and investment banking firms. If you're not looking at those industries, you're not going to be paid as much. Plus, I don't know many safety professionals in those industries.

Call your alma mater's recruiting office and find out what recruiters are offering on campus. Tap into your alma mater's alumni network. It is an old saw that alumni hire alumni. At least I know that's true for Texas A&M University. In Houston there is even a 6:00 a.m. meeting every Wednesday morning for job hunters and those who have been laid off. Their average time to place alumni is six weeks. Check yours out.

If you're thinking of moving from a large company to a smaller one, contact people in venture capital associations who might know about salaries at companies they've backed. Also, look at the Association for Corporate Growth, a group for individuals working at small-growth companies.

Determine Your Value

Judging market demand is critical to any evaluation. Executive recruiters are good barometers: If they're calling you daily, you're in a field hot enough to warrant a premium over survey figures. If they brush you off like too much lint, you're chilly.

Don't just look at job titles, but rather match up your skills and experience levels. Titles can mean different things in different places. A coordinator can mean a field grunt in one place and the corporate guy in another. In recent years, many workers have added extra responsibilities without changing job titles.

Of course, if you find you're undervalued, you shouldn't expect your employer to right all compensation wrongs. If you have leverage (translation: skills that can't easily be replaced and a sterling track record), management will likely make adjustments to keep you. But most companies can only help so much. People who are 25 to 30 percent below market rates may have to jump ship. However, remember this: most people are replaceable.

Before you rush out and claim every dollar you're worth, consider what being in the right corporate culture is worth to you. Is it worth it to move from a creative environment where you have a lot of autonomy to one where you're going to make $3,000 more a year, but people are going to tell you what to do? Consider this carefully.

Strategies for Getting the Raise You Deserve

Here you are, you enjoy your current position, your colleagues, and the direction your company is going. You'd consider changing jobs only for the right opportunity, but the bigger issues right now are, how much are you worth, and are you missing important career growth?

If you want more money and your company isn't forthcoming, you'll have to ask for it. Here's how to do it without creating the image that you are merely a money-monger.

Develop a Tactical Plan

Negotiating comes naturally to very few of us because it generates potential conflict. Even if you negotiate regularly on behalf of your company with contractors, insurers, and the like, the stakes seem much higher when your ego and income are on the line. Preparing yourself well in advance will give you more confidence and a tremendously better game plan than a spontaneous request.

Begin your preparation by asking yourself why you deserve a raise. In our current economy, several external factors may be in your favor. If the stock market is bullish and shareholder and officer compensation is growing, you will have a better chance. Given these two circumstances, it should be much harder to deny a performing employee a reasonable increase. (If the stock market is bearish and shareholder and officer compensation is declining, you will have less of a chance. It will be easier to deny a performing employee even a marginal increase.)

The tight job market also works in your favor. If your company lost you to another firm willing to pay more, chances are you'd be hard to replace. Human Resources and management would have to spend a lot of time and money finding your replacement, not to mention increasing the workloads of others in your absence. Even if your replacement is equally talented, that person would still require some settling-in time (a year or so), which represents an opportunity cost more wisely spent on giving you a raise and reinforcing your loyalty.

There may also be some internal factors in your favor. Have you:

- Saved the company a significant amount of money, like workers' compensation or insurance premiums?
- Increased your department's productivity beyond expectations, like implementing computer technology (such as a video conferencing center)?
- Accepted a lot more responsibility and authority, like assumed environmental or ergonomic responsibilities?

- Raised the visibility of your employer through publications, awards, or professional society involvement? (This may vary depending on the industry where you are employed.)
- Received an outstanding performance review?
- Cost-effectively implemented CMA's Responsible Care™ initiative, ISO 9000, or ISO 14000, allowing your company to compete in the global marketplace?
- Improved your safety and health program to world-class levels?

All of these activities put you in a good position to ask for and get what you want. In fact, they tend to be better ammunition than market forces because they showcase your specific contributions.

Companies are unlikely to make monetary counteroffers to prevent employees from moving to other companies. They'd be more inclined to suggest employees find a way to enhance your responsibilities or develop a better job fit. However, they'd be willing to authorize an unscheduled raise for an individual who has completed some credential, mastered a new skill that would make them more valuable on the job, or made a contribution to the company that goes far beyond their job requirements.

Sell the Benefits

Another critical issue to consider when asking for more money is how your request will benefit your boss and the company. Others are much more likely to help if you appeal to their enlightened self-interest. If you can't answer how your department and company will gain after you receive a raise, and just ask for more money because you think you deserve it, you face a tough sell.

Once you're satisfied that you have good reasons to ask for more money, decide how much to request. Remember that the amount you seek will hinge on what you're worth in the job market as well as your value to your company. Ask trusted colleagues for compensation information or seek out salary data from professional association surveys, career-related Internet sites, the Bureau of Labor Statistics, and newspaper want-ads.

Also, find out your position's compensation range within your company. If you've been performing well, asking for a raise within this range is more likely to get a favorable response than going above it. If you think the internal range doesn't reflect today's competitive job market, tap other criteria for your increase. For example, suppose you just saved your company $200,000 in worker's compensation costs by decreasing the incident rate and the experience modifier. You could make a reasonable case for receiving five to ten percent of the savings.

When you don't have specific guidelines to follow, you'll have to devise a rationale to explain a logical increase. If you can develop a formula that justifies the amount, you're more likely to convince top management. Top management and comptrollers understand numbers. Use them. A word of warning, though: it's very risky to base your request on what Joe or Jane Q. Employee is making. Comparing your compensation unfavorably to your peers sounds like whining. Because companies don't appreciate employees who discuss others' salaries, having this information labels you as a snooping malcontent who's liable to broadcast a sweetened deal to the troops, thus potentially creating a massive morale problem.

If someone came to me to compare his compensation unfavorably with his internal peers, I'd check his performance against theirs. Chances are he doesn't deserve what his peers are making.

Time Your Request for the Maximum Effect

For maximum impact, ask for what you want when:

- You're ensconced in the glory of a great accomplishment. It's hard to turn down a winner who's just done something outstanding for the greater good.
- Headhunters are courting you. Asking for what you want is much easier when you believe you're in high demand.
- The company's profits and stock price are rising rapidly.
- Your request comes before the next budget cycle. It's possible, but seldom easy, to engineer a raise or bonus once the

budget has been finalized and the dollars you are asking for are not included.

Then there's the old way, you evaluate your boss's temperament and approach him or her on an up day. Never underestimate the power of good or bad moods.

Alternatively, think hard about asking for more money if:

- Your work is mediocre. A rising tide doesn't necessarily lift all boats. Individual merit is increasingly important when determining compensation.
- Your company is downsizing. While the economy is thriving, many organizations are in the middle of layoffs. Asking for more money if your company is struggling is futile and clearly ill-advised.
- New management has just arrived. They don't know you from Adam, so from their perspective you have no proven history. They may already have a predilection to replace you with someone from their own inner circle. Therefore, make sure you're on firm ground before you commence the climb.

Another company or internal department has made you a tantalizing offer. In the spirit of fair play, give your management the opportunity to match or exceed it. If you do this with just the right touch of genuine humility, you may get what you want. Be careful though, it can work against you as well. If you don't think you can pull it off, don't try it. If you are genuinely ready to go for more money, then take the offer and leave the company. On the other hand, should you deliver an ultimatum, you may be shown the door. Unless you are that rare individual who is too important to fire, you will have burned that bridge thoroughly. Also, people backed into corners often come out swinging. Even if you get what you want in the short term, you may be making a long-term enemy who'll seek revenge when you least expect it, like layoffs.

Understand the Obstacles

Consider your boss's most likely objections and generate answers to those objections. How will you counter them? Does your boss even have the authority to approve your request? If your boss doesn't have the power to say yes, talk to the person who can.

While formulating your plan, consider your boss's negotiating style. There tends to be two categories of negotiators: flea market hagglers and straightforward players. Both types disdain the other's style, so it's critical to know the personality you're dealing with beforehand. Hagglers love to think they've forced you to leave something on the table. To help them achieve "victory," ask for 110 to 20 percent more than you expect to receive. When they offer 90 percent of your number, you'll both win.

The no-nonsense negotiator wants to know 100 percent of what you want and why. If he makes a counteroffer, be ready with all the responses you can think of to justify your request. You'll probably come to terms a lot faster with him because he wants closure, compared to hagglers who love to play the game.

Before scheduling a meeting to state your case, develop several options. Plan A should be at least 100 percent of what you want. Plan B should be 90 to 95 percent, and Plan C should be something less than Plan B. Each of these options may be cash alone, an increase in your nontaxable benefits, or a combination of the two. Whatever your strategy, be prepared to say why you're interested in each plan and would be willing to agree to a lesser option.

Be Ready to Sell Yourself

With your agenda prepared, it's time to broach the topic with your boss. Start by listening carefully. Most people fail here by thinking about what they are going to say while the other person is talking. The result cam be a disconnect between what your boss is asking you and what your response is, quickly escalating into an argument. Further, your boss may offer alternatives even better than yours. But don't accept or suggest plans you don't want. In the heat of the moment you may be tempted to act impulsively. Don't. Always give yourself a day or two to decide if there is even a Plan D that makes sense.

Look at your conversation as a compromise. Both of you will probably give up something along the way, but each should be pleased by the final outcome. Remember, you'll be working with this person on a daily basis, and he or she has a great deal of control over your career path. Do it wrong and your career path could easily end up on the street rather than the next floor.

If you can't reach a satisfactory agreement, tell your boss that you're disappointed you weren't able to work things out. That's it. Don't threaten your boss or issue an ultimatum. Just act genuinely disappointed and let him or her sweat. They will begin thinking about what you might do. Given the circumstances, you might lose interest in your job, become a morale problem, or move to a new job. Can they live with that? Only time will tell.

Getting Top Dollar: You Only Get What You Negotiate

You are about to change jobs. You've been interviewing and an offer is forthcoming. What is an SH&E professional to do? Well, many of us would accept the first offer that is made, just happy we got one. Before you say yes, remember this is your last bargaining chip before you join the new company.

If both sides really want to make it work, anything is negotiable, from vacation to salary and bonuses. Depending on the job you are seeking, you may be able to request and get things like a signing bonus, a work-at-home telecommuting arrangement, or extra vacation time, just to name a few perks.

But there's a catch, there is no such thing as a free lunch. *You* have to negotiate your salary. You don't have an agent like professional athletes. It is all on your shoulders. No one is going to come up to you and say, "Mark, you are so great. We like you so much. You impressed everyone beyond their wildest expectations, and we want to offer you a *bazillion dollars*!" Pinch me. Wake me up. I must be in a dream! Guess what? You are! Unless your father owns the business, this will rarely, if ever, happen.

You don't get what you deserve, you get what you negotiate. That is, you have to ask for it. You have to ask for it without demanding it.

What do you say? How do you ask for it without really asking for it? You have to be subtle, elusive, humble but confident. You've got to make them think it was their idea, not yours.

Getting the Sweetest Deal

"The company thinks like I do."

One of the major considerations is whether the company has a personality similar to yours. You will be satisfied and most valuable—and last longer and find the experience more beneficial—when you are able to smoothly integrate with the company and possess a comparable temperament.

"I like the people I have met and what I have seen of the working environment."

This is closely related to the previous comment and equally important. Often professionals will choose to ignore the warning signals of what may be an unfavorable work environment because of a higher salary, promises for rapid advancement, or similar factors. You should weigh the people factor very carefully in evaluating the job offer.

"I want a high/low profile job."

Some professionals prefer low profile positions, but you need a high profile job if you want to be promoted. If you are at the center of a corporate activity rather than being confined to an isolated department or geographical area, you will have the opportunity to make contributions to the company's profitability that will be immediately noticed.

"I would be willing to stay on this job for at least four or five years."

You should plan on an acceptable period of tenure for each job. To a majority of companies, four to five years can be considered the right length of time today, not too short or too long (at least until you are over 45 years of age). The extremes in either direction could weaken your employment record. People who do not stay long enough on one job are frequently branded as 'job-hoppers,' and those who stay too long become identified as 'one-company people,' or individuals who do not have the ambition to handle greater responsibilities.

"The compensation/benefits package is what I want."

Many professionals make the mistake of lowering their salary requirement in order to leave a currently bad position, or believe that they can rapidly work themselves back up to their former income level. If the job does not carry the income you feel you need, it is bound to be a source of discontent that will have an adverse effect on your job performance and your sense of well-being. Forget it, unless your immediate monetary needs force you into an acceptance.

"The job isn't likely to cause any family problems."

If the position is one that may require relocation, long hours, evening work, or heavy travel schedule, these factors should be weighed against the probable impact on your family. It is difficult to achieve and maintain a healthy work environment when there are continuing personal problems at home.

"You don't get what you don't ask for."

Employers aren't paid to bestow fat salaries and perks on their new hires. Early in my career, I was transferred from the Houston facility to the Colorado Springs facility. I had to rely on a coworker with twenty-plus years of experience just to know what was available. Once I found out what was available, I asked for everything specifically. Once you get there, it's too late, you have no bargaining position.

When changing employers, the HR people don't get paid to offer you a golden deal. And they won't tell you all the benefits unless you ask very directed questions. The bottom line is, if you don't know what questions to ask, you don't get the benefits.

So how do you find out what the benefits are? Ask your prospective boss, your predecessor, employees (preferably not in the same department or the people you interviewed with), and other people you may know who are employed by the company. They may not want to reveal their own salaries, but they might be gossipy about what others earn. And they're often willing to share what they know about company benefits and typical perks. Once you find out, ask the hiring officials directed questions.

One of the most popular requests involves vacation time. Many companies have rigid vacation accrual policies. This is especially onerous for new employees. Many companies offer the standard two weeks to new hires in their first year, which is quite distasteful to someone like me who has 24 years experience. I have become accustomed to a minimum of three weeks vacation. Some companies can't change their company's vacation policy, but they can bump your salary to accommodate unpaid time off.

Also at the top of the list is work-at-home provisions, or telecommuting. As with other benefits, companies don't volunteer these, but they'll provide them if gently prodded. People are negotiating for more control of their life and schedules. After receiving a job offer you can ask for things like to work from home one day a week, particularly if you have a long commute and can guarantee the quality of your job does not suffer. But be careful, this sometimes translates into the 24-hour job. This can be easier for many safety professionals. I know of many safety professionals who hold corporate slots and work out of their home because they are on the road so much visiting all of the company's regional plants.

"Do your homework and know the facts."

Mark Twain once said, "Get your facts first, and then distort them as much as you please." The best way to get the salary you want from a new employer is to provide ample justification that what you're asking for is fair and reasonable. The Bureau of Labor Statistics (BLS) has a listing of job positions and salaries. A Dun & Bradstreet listing can help you identify profits and may even have salary-related information. Professional societies like ASSE conduct salary surveys. Some even have some salary information, such as the *Industrial Hygiene News*, which does an annual white paper. If you have ever gone to ASSE's website, Nextsteps has a list of jobs and companies interviewing. Look at the positions and salary ranges. Jot them down for later reference. Use whatever information that relates to your job that you can use to justify your salary.

For salary negotiations, nothing beats hard, clear evidence that your price is in line with similar positions and functionalities or with that of other employees in similar jobs at the company. Some search firms keep databases of salary information, which they'll share if you ask. Remember when you bought and sold your house and the price was set based on comparable homes in the area? The same principle applies here. The boom in job hunting via the Internet has also created a growing array of comparative salary data, not to mention a way to ask others about going rates. Surf the net and you may be surprised what pops up. You may even find it on the company's website.

Even newspaper want ads sometimes include salary information. If they don't, call the companies advertising jobs similar to the one you're considering and ask about the salary range.

ASSE's Nexsteps lists jobs on a monthly basis. Many of these jobs actually indicate salary ranges and requirements.

Finding company-specific information requires some extra stealthy investigation. The first place to look is the HR department, which will almost always give you a salary range for a given position. I know some companies that have a notebook at the reception desk that lists all the job requirements and salary ranges. Wow! What a gold mine!

Salaries of top-level executives, of course, are listed in the company's annual proxy statement, which also discloses the perks and other benefits they got the previous year. This can give you a hint to what the safety professionals may earn also. Some public companies will also provide copies of the current senior executives' employment agreements, which are filed with the Securities and Exchange Commission.

It's also important to research a long-standing area of overall offers called "making whole," which basically ensures that new employees don't have to shoulder any expenses when taking a new job. The practice is especially common when job candidates are asked to relocate. But it can require some work to get the package you want. Some companies don't offer relocation. So how do you get it? You will have to convince the HR person that you are worth it, that it is a small investment in the future, that you will be frugal in finding and costing out a move. One other way is finding out that others in the company had their moves paid for. That's leverage.

Even if you're not relocating, it's worth finding out what the relocation budget is for your position. If a given job has a budget of, say, $12,000, you may be able to get them to apply it toward something else. How does this sound? A company car, a home office, or a notebook computer. If you can make a case for something, and the money's available, companies will sometimes go for it.

"Persistence."

Remember, it's always too soon to quit. Vince Lombardi once said, "Winners never quit and quitters never win." When it looks like nothing will happen, keep the faith. If your research or interviews show that your salary goals are out of line with the company's pay ranges, there's still hope. Some companies are providing sign-on bonuses for positions that are particularly competitive.

Beware, though. Companies know that people with hot skills have them over a barrel, and it doesn't help to rub their faces in it. Remember to stay humble. You also run the risk of having your new boss resent that you took him for all you could at the outset. Where your treasure is, there you will also find your heart. If you focus on money and that is all that matters to you, things like loyalty and honor take a back seat. That is not good for the professional who wants a long term position with the company.

It's tough enough starting a new job. You certainly don't want to start it off on the wrong foot by making them resent you because you played hardball over salary.

"Tell the truth—period."

H. Norman Schwarzkopf said, "The truth of the matter is that you always know the right thing to do. The hard part is doing it." It's never a good idea to lie to recruiters. Lies have an odd way of finding their way back to the person that knows the truth. It may hurt and make your resume less attractive, but at least you will be able to sleep at night. You are building your reputation as an SH&E professional on truth, integrity, and honesty. Often it is a small world in the SH&E profession. All you have is your reputation. Protect it well.

If you lie on a job application, it can come back to haunt you. Let's say you bump up your salary on the application by $5,000. Some companies will request a past pay stub as proof of your earnings. Oops! There goes that job opportunity.

"Show them your value before you ask them to show you the money."

Before I even broach the topic of money, I make sure that I demonstrate my abilities and they begin believing that I am the one for the job. No matter what's on your must-have list, the surest way to get the salary you want is to convince the company that you're worth every penny you are asking for. Be prepared to demonstrate your willingness to go the distance during the early stages of the interview process.

I was asked once to meet a prospective employer on Saturday morning or a holiday weekend. I didn't know it at the time, but later found out, that this was a test to see my dedication and willingness to rearrange my schedule and meet them, indicating how I would perform if I got the job. I rearranged my schedule to meet with the prospective employer. In another case, I brought a work product that was similar to a key work product that I would develop during the course of my employment, if I were hired. It got so much attention that they wanted to copy it for later reference. I told them I did not feel comfortable doing that for ethical reasons. From that point on, I had them selling me on the company rather than me selling them on my abilities. The moral of the story is, get them to fall in love with you, then spring your requests on them.

I always bring a portfolio of work products on the interview. I leave them in my briefcase until the right opportunity to present one arises. Be careful though, I have found that unless the timing and opportunity are just right, you can intimidate your prospective employer and inadvertently kick yourself out of the race, rather than jump to the number one candidate.

"Timing is crucial."

Successful negotiations require restraint, and nothing works against you more than giving your prospective employer the impression you're impatient or greedy. As I said earlier, I wait till I know they want me before I bring up salary. I try not to broach the subject of salary until late in the interview process. The best time to do this is when it's clear they're ready to make an offer.

At this point, it's important to get a general sense of the salary range. Most prospective employers are more than willing to provide such information in a good faith effort to make sure everyone's on the same page. But don't be discouraged if their salary range is different from yours. Once you know they want you, it is amazing how soft those hard ceilings get. The bottom line is that the president of the company can do whatever he or she wants, given the right candidate for the job.

You need to listen for leading questions—if they make an offer and then ask, "How does that sound to you?" or "Tell me how you feel about that," you know there's some flexibility. Never, ever, jump and say. "Yes! Yes! Yes!" no matter how good the offer is. You don't want to sound too eager. Tell them that you have to speak with your husband or wife and will get back to them tomorrow. One employer made an offer during my closing presentation in the interview and wanted an answer on the spot. I had to refrain from jumping at the chance to say, "Yes! Yes! Yes!" Prospective employers usually want a response right away. But you should avoid this pressure. One of the biggest mistakes people make is to accept an offer too quickly. Sure, you want to show enthusiasm, but there's no need to be impetuous. Be polite and defer them for at least 24 hours. This is almost always an acceptable response to any job offer. Besides, being a little elusive never hurts. You want to keep them jumping.

Another aspect of timing is interviewing to get several offers within a short period of time so that you don't have to stall too long to maximize your best opportunity. Once, while I was interviewing, I was waiting on three offers. I wanted out of my current employer really badly. Two of the three came within two days of each other. The third was delayed. When they called I informed them that I already had two offers and wanted to take their offer into consideration. Since I had made it through two of the three interviews they had scheduled, they jumped through hoops to make an offer. I was able to evaluate three offers within several days of each other, rather than having only one offer to evaluate. That way you can make a comparative evaluation.

"The Counteroffer."

This is a difficult call. Considering a counteroffer depends on the company culture, your reasons for leaving, and your relationship with your employer and coworkers. As a rule, if I have decided to leave my employer, money is only one element of the whole job picture. As a result, I rarely consider counteroffers. The dilemmas and game playing can really make your head hurt if you aren't prepared.

For instance, I was about to leave one employer and just as the moving van showed up, my current employer called with a counteroffer. I was compelled to send the moving van away so that I could consider the offer. I ultimately decided to take the original offer and turned down the counter. The reason was that I had decided to leave, and money was only one element of the overall job picture. In hindsight, I would have declined the counter offer right away, loaded the moving van, and left for my new employer.

However, let's consider another example. Let's say you are laid off for financial reasons. You find a temporary job at less pay. The company is a good company, but can only pay what they offered you as a temporary. Within a month or so you get called back to your former employer. You then return to your former employer at your exit salary. Within a month or so you get an offer from the company you had a temporary job with, offering enough to leave. You inform your current employer and they promptly counter the offer with a better salary. What do you do? Well, there are many factors, as I stated earlier. If they don't measure up, it is time to leave. If not you may want to stay.

In the final analysis, unless there are extenuating circumstances, I turn down counteroffers. But every situation needs to be evaluated on its own merit. As I said, it can be a difficult call.

These are just a few of the strategies to maximize your salary during a job change when the market is hot. Remember, you don't get what you deserve, you get what you negotiate. Careful negotiation can ensure that you maximize your job change, counteroffer or not, so that you won't be looking for another job very soon after you start your new one.

At the end of the day you want to be able to feel that you got the best deal of all the options that were placed in front of you. In the words of H. Jackson Brown, Jr., "Success is getting what you want. Happiness is liking what you get." Or even better, Logan Pearsall Smith said, "There are two things to aim at in life: first, to get what you want, and, after that, to enjoy it. Only the wisest of mankind achieve the second."

It's also crucial to bring up any nonnegotiable points early on. For example, communicate location restrictions, like not wanting to live on the East Coast or West Coast or any other particular region.

When the time finally comes to dicker over salary, you'll know whether there's room to move, in part, by who presents the offer. Typically, smaller companies without Human Resources departments are more flexible. If they are larger companies, try to deal with the VP of Human Resources.

You need to listen for leading questions. If they make an offer and then ask, "How does that sound to you?" or, "Tell me how you feel about that," you know there's some flexibility.

Hiring managers usually want a response right away. But you should avoid such pressure. One of the biggest mistakes people make is to accept an offer too quickly.

Sure, you want to show enthusiasm. But, as I stated earlier, there's no need to be impetuous. A polite, "Can I sleep on that?" is almost always an acceptable response to any job offer. Besides, a little bit of coyness never hurts, so long as you don't leave your suitor at the altar.

How to Bypass Your Firm's Salary Structure and Negotiate a 35 Percent Raise

I've noticed that most people have no idea of how to gain rapid salary increases or rapid advancement. Yet there are simple approaches that can accomplish these things with surprising speed.

There's no one magic question that, when asked, will always make your firm start writing you much bigger checks. My experience has shown that seeking the kind of advancement we're talking about here rarely works for newcomers. You have to be ready, with your political network strongly in place.

How Not to Get Ahead Fast

Most people think in a *linear* fashion, believing that promotions will come as they move upward in a line from position to position. This system works, but slowly. Other people have more developed theories on advancement, believing that throwing all their effort at building departmental profits or something else will gain the attention of upper management and result in their advancement.

This system is really only a variation of linear thinking and it's not much more effective.

The Key to Advancing Rapidly

In today's work climate, the key is to look outside your firm even if you intend to remain there. Look at trends in your industry, look at how your firm is conforming to them or failing to and position yourself several steps ahead of the field.

The bottom line is this, determine the position your company is going to need to have filled in six months or a year and then *hire yourself for the job* before it is actually created.

New geographical areas. If you know a lot about an area of the country, or even a foreign country where your firm has not yet made inroads, you may have a golden opportunity at hand. Especially valuable: an insider's knowledge of a foreign country or fluency in a foreign language. (Note: Much of American business still remains phobic about foreign languages; you have a great opportunity to position yourself favorably if you know one.)

New joint ventures. Consider the nature of your company's products or services, and the possible benefits of a joint venture with another business.

Conclusions

In negotiating your salary, you need to know what you are worth in your business sector, you need to be a performer and you need a little savvy. As I have stated repeatedly, you need a plan, you need to execute that plan, and be ready to leverage the situation as it presents itself. Timing is crucial, so your plan needs to be flexible for various opportunities. Demonstrating your value before you take the position can change you to a known quantity with a proven track record. It doesn't end at the counteroffer. You can really gain headway, but you need to be careful to not alienate your hiring official. In the end, you can maximize your opportunity to negotiate the best offer. After all, that's what we are all working for anyway, isn't it?

Success Tips

How to Find Out What You're Worth

1. *Industrial Safety & Hygiene News*
2. *Occupational Hazards*
3. *Occupational Safety & Health*
4. *Business Week*
5. *Money*

Determine Your Value

1. Headhunters
2. Job titles and experience levels

Goal Setting and Time Management

Avoiding the Busy-ness of SH&E

As stated throughout this book, we all need a plan to get through our career. Without a plan we are just floating along. We get caught up in other people's priorities and never accomplish anything meaningful because we have no plan. Likewise, why worry about time management if we have no goals? So let's talk about goal setting and then get to time management. This will allow you to focus on things that bring reward, accomplish tasks in a timely manner and achieve your goals throughout your career.

Goal Setting

It has been said that without goals, any road will do. It seems that people today can tell you what they would do if they won the lottery, but they can't articulate where they want to be in five years. Without a plan that can be executed to drive your career, you are merely floating to retirement based on someone else's priorities. If you fail, it wasn't your fault, it was someone else's plan. In the end, people don't plan to fail, they fail to plan.

A five-year plan is a good place to start. Think about where you want to be and how to get there. In 1990, I decided to exit the aerospace industry for the oil and gas industry. I put together a five-year plan and executed it. My first goal was to find a transitional company that was a half-step out of aerospace, which I successfully executed one and a half years into the plan. I then found a petrochemical processing company where I stayed for two years. The process was complete when I moved to the largest independent oil and gas operator in the country by 1996. But it could have never been successful without a plan.

So, I encourage you to think on where you are and where you want to be. Then devise a strategy of how to get there. Before you know it you will find that you have accomplished your goal. It's just that simple. It is not rocket science.

Goal Setting Rules

Here are a few rules to think about when you formulate your plan.

Your goals must be in alignment with each other. There must be harmony and synchronization of your goals. For example, attaining a master's degree in safety matches up with attaining more responsibility in your job. However, attaining a liberal arts degree in dance does not match up.

Your goals must be achievable. One fault of those who set goals is setting them too aggressively, making them unachievable. When you do this, you get discouraged. The acid test I use is, "Does this goal sound like I am trying to become a jet pilot?" This would be a goal that I could never achieve due to many physical attributes. If the goal sounds like this, I back off. The best way is to set many small achievable goals. One of the things I tell those asking is that every step I take in some way takes me closer to my five-year goal.

Your goals should have both qualitative and quantitative measures. Qualitative measures are those that improve your stature in whatever your pursuits are, for example, achieving a professional designation like a CSP or an improved job title. Quantitative goals include achieving a certain salary level or financial investment figure, anything that can be measured with a number.

I would hope that all of your goals would have a positive effect on your family. For example, getting a six-figure salary that requires you to travel 50 percent of the time would definitely have a negative effect, while the same salary and ten percent travel would have a positive effect on your family. Any job that has less travel with an increase in salary would have a net positive effect on your family.

Your goals need to be long- and short-term. Short-term goals are the small steps I was speaking about before. Each one takes you closer to your long-term goal. Just to remind you, long- term goals are not winning your state lottery.

Think deeply about what your goals are. Do some soul searching, so when you have completed setting out your goals, you have a plan and a methodology to achieve them.

Best of all, when you review your goals annually, you can perform midcourse corrections. If you find that you are way off your path, you may choose to do something radical like change jobs or professions. The decision becomes much simpler when you have established a plan.

How Do You Identify Your Goals?

What are the five most important values to life? Just to give you an example, here are mine. Don't feel like yours have to be the same, but you need to know your values when setting your goals.

1. Spiritual
2. Family
3. Work
4. Financial security
5. Peace

What are the five most important goals in your life?

Here is a good start—ask yourself the following questions:

- If you could do anything that you wanted to do, and failure was not an option, what would you do?
- If you had a million dollars, what would you do?
- What do you most enjoy doing in life?

Ten-Step System to Reach Your Goals

Follow these steps to help you reach your goals.

1. Develop a burning desire.

Without such desire, you cannot achieve your goals. You must have a dream. Martin Luther King did not galvanize a generation with, "I have an objective," he had a *dream*. This connotes passion. You must have a burning in your gut to achieve something and pursue it with all your passion. If you don't have this passion for what you do, quit and go find something for which you do have a passion.

2. Develop a belief.

This is the well you draw from when you are thirsty, and it is the well that keeps you going. Whatever it is, you must have an undying belief that you can cling to. For those who do not, I offer the fact that those who stand for nothing will fall for anything.

3. Write it down.

There is an old saying in the insurance industry, "If it isn't in the file, it didn't happen." If you don't write it down and review it every so often, how do you know where you are going?

4. Analyze your starting point.

Where are you today? Think about doing a gap analysis and seeing where you are today and where you want to be, and start from there.

5. Set up deadlines.

If you have no deadlines, you will put it off until it finally falls off the to-do chart. Deadlines force you to get off of square one and do what you said you wanted to achieve. It is a bit of self-accountability to follow through with your plan.

6. Make a list of your obstacles.

Which ones are obvious so you can plan? This includes getting a degree, a certification, or a transfer to a different position to department. There are obstacles that you can't list—they are the ones you will just have to deal with as they come at you.

7. Compile information that you will need to complete the trip.

What are your information resources that you will draw from to make sure you are going down the right path?

8. Make a list of people that you will ask for help.

I know many men abhor asking for directions, but it is useful to ask a mentor who has been down the path and made a few mistakes, so you don't repeat them.

9. Develop a plan.

People plan their vacation more than they plan their life. If you don't, someone else will. I've already beat this dead horse enough.

10. Make a decision to succeed.

Determine that there is no turning back. A book I recommend to some employees is *The Go-Getter: A Story That Tells You How to Be One* by Peter Bernard Kyne. This book tells you everything when it comes to going through your obstacles rather than stopping and quitting at every hurdle.

How can you overcome these obstacles?

- Make an "everything you've ever wanted" list.
- Evaluate where you are today, physically, mentally, emotionally, and spiritually.
- Describe the perfect you.
- Do a comparison of now and the future. Keep a journal.
- Set small goals.
- Reward yourself for your successes. Celebrating gives you the energy when you have none later on.

EAT Philosophy

The EAT philosophy is from Bryan Dodge, a leadership and management speaker. These materials and many others are available from his website (www.bryandodge.com).

E stands for **Eager learner**. Your mind has to be growing. Learn about everything you want to know. If you are learning, your brain is active; if not, it is dormant.

A stands for **Argue** or debate. You can either ignore issues or resolve them. But do it on its time. You now have control of the issue. If you ignore issues, they get worse and usually stress you out. The number one killer in the world today is stress.

T stands for **Thankful**. Many say the grass is greener on the other side of the pasture. Why? Because we keep looking over there. In reality there is more manure over there than you think. I always tell employees who are about to leave, pick your bosses carefully and remember you are merely exchanging one set of warts for another, and every company has them.

Focus on things you can control.

These are questions all successful people answer in the same way:

How many books have you read this past year? Readers are leaders. CEOs read five books a week—four are technical, one for entertainment.

How many classes or seminars have you taken this past year? Keep learning. Taking classes gives you skills. No new skills, no new value. The easier you are on yourself, the harder life will be on you. The harder you are on yourself, the easier life will be on you. Did you know that you could earn the equivalent of a master's degree by listening to tapes on your 30-minute commute to work every day rather than listening to the mindless drivel of talk radio? Think about it.

How much money have you saved and invested this year? Hire your money. Make it go to work for you. Don't fire it by spending it, especially on stuff you don't really need. You have to pay yourself first.

The Key Components to Greatness in Life

The key components for greatness in life is also from Bryan Dodge. These materials and many others are available on his website (www.bryandodge.com).

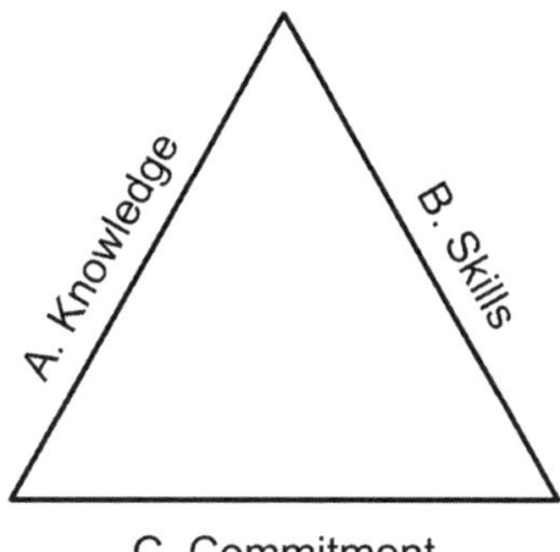

A–Knowledge. The information required to do your job. People who claim they have learned everything that they want to learn are dangerous and not motivated.

B–Skills. The tools you need to do your job. If you are learning, you are on the skill side of the triangle. You will be more focused and committed to your goals. As a result, your knowledge will grow.

C–Commitment. The key is commitment. If you need an alarm clock to wake up you are not committed, you are on automatic pilot. If you are committed, you can't wait to get up to go to work.

If you have these three components, there is a fourth that determines just how successful you will be. Can you identify what it is?

What Am I?

I'm your greatest friend or your heaviest burden.
I can carry you on to success or tear you down to failure.
And yet I am at your complete control.
Half of the things you do, you turn them over to me.
And I can do them effectively and efficiently.
I am easily managed; you merely have to be firm with me.
I'm the servant of all great men and women.

I'm the master of failure.
I'm not a machine, but I work with the efficiency of a machine.
And the intelligence of the human body.
Run me for profit or turn me for ruin, makes no difference to me.
Be easy with me and I'll destroy you.
What am I?

And the answer is: Discipline.

Setting Goals for Employees

If you want your employees to improve productivity and morale, just ask them to do it. That's easy enough. Right? Wrong! A surprising number of managers have trouble explaining what they want done and why. Here are a few tips for managers to help employees set goals that they can reach.

Assign clear, specific, realistic and useful goals. Goals like, "just do your best" can't replace specific goals. Also, goals in the form of sound bites are terribly ineffective. The more specific your stated goals, the more credibility you gain.

Be a positive performance role model. A manager's positive expectations often set the stage for higher performance and create a positive association among you, the performer, and the organization's success.

After setting goals, employees should develop their own way to achieve them. Personal responsibility for developing strategies to reach goals makes managers look secure.

Provide specific feedback. To improve employee's performance and develop a positive impression of yourself in their eyes, give timely feedback about reaching the goal. You will be seen by them as involved, informed, and helpful when you offer specific knowledge of results to your people.

Offer fair rewards for achievement. Providing appropriate levels of recognition for your people will help you be viewed as fair. Management fairness is always important for employee morale and satisfaction, while strengthening the goal-setting process.

Adjust goals as new information is available. Flexibility is crucial to an effective manager. However, changing goals too frequently can make managers appear too unfocused.

Follow up regularly on performance targets. Regularly track progress toward goal achievement and discuss this process with your employees. A manager then will be perceived as being on top of the situation. You can risk losing credibility with your employees if you fail to follow up.

So, if you want your employees to improve productivity and morale, just help them set goals using these principles and watch the achievement overwhelm you.

Never say this to your boss about your employees:

- "They didn't get back to me," or "They're getting back to me." Both tell your boss that you lack the initiative to reach them.
- "I thought someone else was taking care of that." The message: You focus on your small tasks and ignore the big picture.
- "No one ever told me." Your boss will think you're out of the loop, oblivious to what's going on around you.
- "I didn't think to ask that." This shows the inability to see down the road and make connections between one problem and the next.

Time Control and Management

It seems that, without managing our time, time has control over us. We never have time to do the things that need to get done. The key time control is getting time to work for you rather than you working for it. Time is something of which everyone has the same amount. So those who accomplish great things obviously manage their time well. Now, so can you. Just remember, every time you lose a minute, you can't replace it.

Your life, your income, and your results are determined by the way you control your time. What you spend your time on should bring the greatest return affecting your income. If you spend your time on the things that matter, you will get results that subsequently affect your income. It sounds logical.

Plan your time strategically to give yourself the highest return on your energy. Don't waste your time on things that have no return on your investment, but rather those that add value to your goals.

It's all about priorities. Time managers know how to set them. Think of it this way: practical people know how to get what they want. Philosophers know what they ought to want. Leaders know how to get what they ought to want. In the end, success is the progressive realization of a predetermined goal.

What is required for you in your job? What gives you the greatest return? What is the most rewarding? There are typically three common problems:

- Abuse—too few employees are doing too much
- Disuse—too many employees are doing too little
- Misuse—too many employees are doing the wrong things

Priorities never stay put. They are constantly changing, so we need to know where to put our effort and attention. Consider the following:

- Evaluate—3 Rs (Requirements—Return—Reward)
- Eliminate—what am I doing that can be done by someone else?
- Estimate—what are the top projects this month and how long will they take?

You cannot overestimate the unimportance of practically everything. This bears repeating. Ask yourself:

- If you had to do things over, what would you do differently?
- If I had to do it over again, would I reflect more?
- If I had to do it over again, would I risk more?
- If I had to do it over again, would I do more things that would live on after I am gone?

Applying these questions may severely affect what we are doing and significantly pare down what we are doing to what is really important and what is a waste of time.

Good is the enemy of the best. How do you break the tie between two good options? Here are a few thoughts. Ask your boss or coworkers their preference. Can one of the options be handled by someone else? Which option would be of more benefit to the customer? Make your decision based on the purpose of the organization.

You can't have it all. I know we live in a world that tells us you can every day. Just listen to the car commercials, or any commercial for that matter. We need to understand that we need to focus on a few priorities. If we don't, too many priorities paralyze us, and we do nothing because we think it is too much to handle. Further, when little priorities demand too much of us, big problems arise and we focus on the little things rather than the big things, to our demise. Implementing time deadlines force us to prioritize.

For example, Parkinson's Law states, "Work expands so as to fill the time available for its completion." For example, if you have one letter to write it will take all day to do it. If you have 20 letters to write, you'll get them done in one day. In the end, too often we learn too late what is *really* important. There are no tombstones that say, "I wish I had spent more time at work."

The Pareto Principle, when applied to time management, states, "Twenty percent of your priorities will give you 80 percent of your production, IF you spend your time, energy, money, and personnel on the top 20 percent of your priorities." This is the 80–20 rule that you often hear mentioned in conversations. Here's an application.

Time: 20% of our time produces 80% of the results

Counseling: 20% of the people take up 80% of our time

Products: 20% of the products bring 80% of the profit

Reading: 20% of the book contains 80% of the content

Job: 20% of our work gives us 80% of our satisfaction

Speech: 20% of the speech will produce 80% of the impact

Donations: 20% of the people will give 80% of the money
Leadership: 20% of the people will make 80% of the decisions
Picnics: 20% of people will eat 80% of the food

Choose or Lose

When managing your time, you can be a leader and choose to do what is important, or lose and become a follower or an also-ran.

Leaders:	Followers:
Initiate	React
Lead	Listen
Pick up the phone and make contact	Wait for the phone to ring
Spend time planning	Spend time living day-to-day
Anticipate problems	React to problems
Invest time with people	Spend time with people
Fill the calendar by priorities	Fill the calendar by requests

Take the long view. Have a vision and do things today for tomorrow.

The only thing separating where you are today and where you want to be in twenty years is sacrifice. Similarly, there are two types of pain: discipline weighs ounces and regret weighs tons. I can recall going to work at Ford Aerospace in Houston for NASA/Johnson Space Center in the mid 1980s. This was when the space shuttle was coming into the picture. I was listening to a conversation that had an impact on me the rest of my life. It was a statement made by a fellow who was in his mid-50s. He said, "You know, if I had only struck out on my own ten years ago, I could have retired today with what I know right now." That statement struck home with me. I vowed that when

I was in my mid-50s I would not be able to say that, because I lived my career with fervor and sacrifice and discipline, not regret and "woulda, coulda, shoulda".

Along the way, remember to keep your life in balance. Yes, you can have balance in today's microwave "got to have it now" mentality. Here is another unconventional principle: it's *quantity* of time at home and it's *quality* of time at work. When you are at work, don't spend your time gossiping, talking about something irrelevant, reading the newspaper, or playing video games. WORK! You are working so you can get home and spend time with your family. You are working to live, not living to work.

Here is the Golden Rule activity test: When I am at work, is what I'm doing in direct line with my goals? When it isn't, you are wasting time that could be spent at home. At the risk of being redundant, every action and activity at work is a step towards goals that you want to accomplish. Set your priorities and stick to them.

Defining "Urgent"

In his book, *The Tyranny of the Urgent,* Charles Hummel divides urgent and non-urgent tasks into two categories, as follows:

	Urgent	**Non-urgent**
Important	I. Important–Urgent	II. Important–Non-urgent
Not Important	III. Not Important–Urgent	IV. Not Important–Non-urgent

He distinguishes urgent from important tasks as follows:

- Urgent: Things that have the appearance of demanding our time.
- Important: This has to do with results and how you measure them.

Here are tips for managing your tasks:

- Write a list of things you are going to do for the next day and leave it at work. If you do, you are in control when you write them down.
- When you wake up, get up. If you don't, you will get up more tired than when you first woke up.
- Stay on each item until completed, regardless of interruptions. Don't let the time dictate which item you are working on. Finish the task. Do not go to the next item until the one you are working on is completed. Check it off.
- Don't take work home. If you cannot get the job done in the time allotted, you do not have the skills to do the job.
- Most of us spend our time in **Important–Urgent**. We put a task off so long that it becomes Urgent. What happens is that the quality of our efforts is compromised. Set your deadlines. If you are given 30 days, set a goal of 15 days.
- Spend most of your time working in **Important–Non-Urgent**. Get a task done before it becomes Urgent.
- SED a Not Important–Urgent task: **S**implify it, **E**liminate it, or **D**elegate it. It is the biggest time robber we face today.
- A **Not Important–Not Urgent** task means downtime. Relax. Go there because you want to, not because you have to.

What Is Your Time Worth? What is Your Return on Investment (ROI)?

You need to know what your time is worth. Evaluate each task according to its time cost and refuse to spend valuable time on a worthless task.

If your annual earnings are:	Every hour is worth:
$12,000	$6.25
$15,000	$7.68
$18,000	$9.22
$20,000	$10.25
$25,000	$12.81
$30,000	$15.37
$40,000	$20.49
$50,000	$25.61
$75,000	$38.42
$100,000	$51.23
$200,000	$100.45

So, now you have all the information, but no tools to perform time management. On the next page I have provided a daily to-do list that can be used to formulate your day, week, month, and yearly planning tasks. What I recommend you do is, at the end of the day, take five minutes and fill it out for tomorrow and (I know this is sacrilege) leave it at work for the next day. It contains:

- Phone calls to make
- Personal notes
- Things to do *now* (High Importance–High Urgency)
- Things to do (High Importance–Low Urgency)
- Things to delegate (Low Importance–High Urgency)

Using this tool will make your day go smoother and more efficiently and, hopefully, less stressfully. I have used it for years and it works for me.

Date ___/___/___

Phone Calls to Make		
1		☐
2		☐
3		☐
4		☐
5		☐
6		☐

Personal Notes		
1		☐
2		☐
3		☐
4		☐
5		☐
6		☐

List of Things to Do

(High Importance–Low Urgency)		
1		☐
2		☐
3		☐
4		☐
5		☐
6		☐

List of Things to Delegate

(Low Importance–High Urgency)		
1		☐
2		☐
3		☐
4		☐
5		☐
6		☐

List of Things To Do Today
(High Priority–High Urgency)

Priority		Completed
1		☐
2		☐
3		☐
4		☐
5		☐
6		☐
7		☐
8		☐
9		☐
10		☐

Secrets for Success and Getting Things Done

Concentrate

When you concentrate on a task, you focus your attention on it and that, in turn, focuses your energy. It's hard to be distracted when you're truly concentrating. When I am very busy, I don't answer every phone call unless I know it is an emergency. I check my messages at 10–2–4 (10 a.m., 2 p.m., and 4 p.m.). I don't return all calls. I selectively answer the business calls when I know I will be leaving a message and leave a detailed message so that I will not need a return phone call. This allows me more time to concentrate and focus on my immediate emergencies.

Honor Your Personal Time Clock

Are you a morning person who jumps out of bed at 5 a.m. rarin' to go? Or are you a nighttime person who can't function before 10 a.m. and hits your stride at 2 p.m.? If you're a morning person, schedule presentations and make important decisions early in the morning while you're in your prime. If you're a nighttime person, schedule important activities in the afternoon or early evening. If you have an early morning meeting, prepare everything you need the night before, and eat a high protein breakfast.

If you don't honor your time clock, you will find that you will run out of gas in the middle of the week without time to recharge your batteries. It would not reflect well on you if this occurred in the middle of a big meeting with the plant manager or the president of the company.

Motivate Yourself and Others to Take Action

People are either motivated toward a goal or away from the consequences of not acting. People are motivated for a wide variety of reasons. When you understand other people's motivational styles, you can communicate goals and performance requirements so that people respond in the way you want them to.

Know When to Act

I don't respond to everything that comes across my desk. I ask three questions about every activity: Does the activity have value? Does the activity have a deadline? Is the activity urgent or does it have a deadline of less than one week away?

When I ask the question, "Does the activity have value?" I don't do things just because I have always done them. I try to think outside the box to determine if there is a better way to do it. If there is, I pursue doing it that way rather than the old way. I further analyze if we need to even do it at all. If I don't need to do it at all, I don't do it anymore, period. Secondly, I want to make sure that I am improving the process rather than taking away from the process. I don't touch anything if I cannot add value to it.

When I ask the question, “Does the activity have a deadline?” I want to know when I should schedule the activity: today, tomorrow, the end of the week, two weeks, next month, next quarter, or next year. Just because it came across my desk today does not mean it has to be done today. If I have time, I’ll do it. But if other things more important are pending, those items take priority. If it is due more than two weeks away, I put it on my near-term calendar.

When I ask the question, “Is the activity urgent or does it have a deadline of less than one week away?” and I determine it is immediate or within the two week window, I can schedule it on my short-term calendar.

Work Smarter Every Day

I try to learn something every day. That is the kind of person I want to be.

Secondly, I try to improve my communications skills every day. Fifty percent of all communications fail. I try to repeat myself whenever necessary and try to make my first communication clear and within the attention span of the hearer.

These are just a few of the secrets I use to get things done. Remember, we never have time to do it right, but you always have time to do it over.

Conclusions

Avoiding the “busy-ness” of dealing with many tasks but accomplishing little requires discipline and focus. It also requires spending time on things that matter and not on things that do not. If you use the table from *The Tyranny of the Urgent* in this chapter and the daily planner, perhaps you will now be able to accomplish more. In the same manner, you will also be able to reduce your daily stress and enjoy life outside of work while accomplishing your life’s goals. Having goals, managing your time, and focusing will make you a better, stronger, more efficient SH&E professional.

Success Tips

Goal Setting

1. Develop a Five-Year Plan
    - Goals must be achieveable.
    - Goals must be qualitative and quantitative.
    - Goals must have a positive impact on your family.
    - Goals must be long- and short-term.
2. Setting Goals for Employees
    - Assign clear, specific, realistic and useful goals.
    - Be a positive performance role model.
    - After setting goals, employees should develop their own way to achieve them.
    - Provide specific feedback.
    - Offer fair rewards for achievement.
    - Adjust goals as new information is available.
    - Follow up regularly on performance targets.

On Management and Leadership

Finding, Recruiting, and Keeping Excellent Staff

In today's job market, it can be difficult to recruit and retain top-notch employees. That means you have to be more proactive and more creative in your approach to finding and keeping staff.

With today's competitive market it's not crystal clear which task is harder: recruiting or retaining staff. Today's job market often resembles an auction in which you bid for the people you want to hire and hope that they accept the package.

Companies are learning to live with the message they sent during the last several years: "Don't expect a guaranteed job for life. You're responsible for your own career." Employees are acting on that message. They're changing jobs frequently enough to live up to the estimate that individuals will typically work for six companies before their career is over. (Some estimates run as high as 11.)

Retaining employees is now a bottom-line proposition for American business. The cost of finding, hiring, and training a new employee equals six months of annual salary, and it continues to increase.

Dealing with the Challenge

A number of companies are implementing a two-part program to deal with this critical human resources issue. One part involves the use of technology as a recruiting tool, and the other involves the implementation of a proactive policy to make the workplace flexible and employee-friendly.

For starters, companies are not limiting their recruiting to the traditional sources of new hires such as advertisements, search firms, and networking. They are also posting jobs on the Internet, which has become the de facto source for both job seekers and employers alike.

Given the dramatic increase in electronic recruiting, some predict that this trend is just beginning in response to what has been described as "a permanent labor shortage" in the United States. To facilitate the recruiting process, technology is providing tools for screening potential recruits and pinpointing potential hires.

An example is Futurestep (www.futurestep.com) from the global recruiting firm Korn/Ferry, which is headquartered in Sherman Oaks, California. Futurestep uses the Internet to register candidates online and match them to client openings for mid-level manager positions.

Futurestep uses a series of website questionnaires to assess work styles, motivational drive, qualifications, and experience. This helps search professionals select prospects for the next step: videotaped interviews that are eventually sent to clients. The goal is closure within 30 days, compared with the traditional four-to six-month search to fill a position.

To succeed in the hiring stage, many companies are becoming more flexible and innovative. Some are offering three weeks of vacation in the first year, stock options for employees below executive levels, company cars, signing bonuses, company-paid meals, recreation rooms, four-day work weeks, and company-paid training.

Some of the enterprises with the most success in competing for talent have streamlined the interview process so they can hire people quickly, before another firm makes them a job offer. In one approach, each company executive who interviews applicants is limited to one area, such as technology, leadership ability, or interpersonal skills, rather than having all interviewers cover all qualifications.

One company even had a hiring weekend to screen short-listed prospects. As a result, management was able to offer candidates jobs the following Monday morning, while also saving $200,000 in recruiting expenses.

Nevertheless, the talent hunt doesn't stop once a company signs someone up. Management also needs to keep that person on board. What attracts people to a company in the first place—salary and benefits—is not what builds commitment. Once hired, employees view salaries and benefits as entitlements. Therefore, managers must have the flexibility to adjust working conditions to match the different needs of their staff members.

For one thing, managers should consider differences in life situations. After all, single people in their twenties have different needs and wants than parents with young children or older workers who have kids in college. For some employees, flextime is a big plus, while others favor telecommuting. For many, company-sponsored training is a real motivator.

As a result, you need to take a proactive approach to building employee commitment. You need to assess the most important factors: recognition of the importance of personal and family time; connection with the company's vision; opportunities for personal growth; ability to challenge the status quo; and satisfaction from everyday work. Employers can build employee commitment to the company, but you have to earn that commitment with policies that meet the changing needs of their employees. Otherwise, you may find yourself beginning the talent hunt all over again.

Communicating with Your Subordinates

When the word *leader* is mentioned, a certain archetype comes to mind. At the office, it's the polished, professional, handsomely-tailored go-getter who works 80 hours a week, never tires, never sweats, and never entertains doubt or uncertainty. In front of subordinates, you're the charismatic, chin-in-the-wind leader who mesmerizes or motivates through the courage of your convictions, the strength of your ideas, and, not incidentally, a shrewd use of Aristotle's rhetorical tool kit. Your personality is strong, your vision bold, and your spirit indestructible. These are the characteristics of leadership we are taught to admire and emulate, nurture, and develop. But are these really the character traits and personal styles exemplified by true leaders in today's professional environment?

According to some of the latest research on the characteristics of effective leadership, the answer may very well be no. Just as the definition of good corporate leadership has evolved in recent years, so too has the thinking on what it means to lead effectively. The rash of ethical high jinks perpetrated by corporate America's top leaders has made employees more skeptical than ever about communication emanating from management. These days, a dynamic leader is often more likely to raise the alarm that the messenger is a salesman or charlatan than to instill trust and faith in one's leadership abilities. This new research, coupled with the well-publicized greed and malfeasance of numerous upper executives, has led many companies to rethink their own leadership models and processes for hiring and promoting people into the management ranks.

New-school Leadership Models

Two significant leadership studies have helped spur this re-evaluation. In his landmark research and bestselling book, *Emotional Intelligence*, Daniel Goleman suggested that the quality of emotional intelligence in leaders—a combination of psychological, interpersonal, and self-management skills—plays a far greater role than IQ or technical skill in their ultimate success. More recently, in the best-selling book *Good to Great: Why Some Companies Make the Leap...and Others Don't*, author Jim Collins and his research team found that many of the "great" companies that sustained extraordinary long-term results featured distinctively different kinds of leaders. Dubbed "Level 5 leaders" to denote the top rung of a leadership hierarchy, these executives are the antithesis of the egocentric, celebrity leader often championed by the business press. Level 5 leaders feature a unique blend of personal humility and professional will, Collins found, and are more "plow horse than show horse." They are largely self-effacing individuals who have almost a stoic resolve to do whatever it takes to make a company great. Although they're not devoid of ego or self-interest, according to the researchers, their ambition is first and foremost for the organization's well-being, not for personal glory.

Surprisingly, Collins believes there's no shortage of Level 5 leaders in most organizations, it's just that they're not always easy to find. What are the telltale signs of the elusive leaders that may walk among us? "Look for some part of the organization where extraordinary results have been produced but no one has stepped forward to take excessive credit for them," says Collins.

Most leadership researchers agree that projecting an aura of leadership will always be about establishing a strong physical presence, having a command of the discipline, and using language that engages and inspires rather than deflates or sedates. But the research also suggests that people are increasingly looking for the presence of other skills and almost ineffable attributes when making the often-difficult decision to believe in, trust, and follow someone from a position of power. Chief among those traits include the willingness to share hard truths without losing a sense of optimism, congruency between what they say and what they do, and an unwavering belief in their vision. Also, a self-awareness indicating that leaders don't see themselves as an unassailable expert, but rather a fellow human being, flawed and imperfect like everyone else, and not too conceited or insecure to admit it. Leaders who don't check at least some of their ego at the door are asking for trouble.

More self-aware, adaptive, humble, and results-oriented leaders are who many leadership experts believe offer organizations better results over the long haul, and who also provide lessons on how best to lead others.

A Quiet Persuasion

So-called Level 5 or high EQ (emotional quotient) leaders tend to communicate strong professional knowledge, an aura of authenticity and "we're all in this together" attitude is usually more persuasive in the trenches than anything Tom Peters or Tony Robbins might say.

Most Level 5s also possess a talent for distilling complex messages into a few key phrases or words so that everyone can understand them.

Jim Kouzes, co-author of *The Leadership Challenge* and *Credibility: How Leaders Gain and Lose It*, is a pre-eminent leadership researcher who has studied corporate leaders for more than 20 years. If there's a recurring theme from the research he and colleague Barry Posner have done, it is this: credibility is the foundation of leadership, and without it, efforts to lead are usually doomed to failure.

"People won't believe the message if they don't believe in the messenger," says Kouzes. "Credibility essentially is doing what you say you're going to do. You can't stand up in front of people and announce policies you're not willing to personally implement, or make promises you know in your heart you can't keep."

Consistency between what you say and what you do can affect credibility, since people don't need to be downwind to sniff out a phony. I watched one executive deliver a really wonderful, upbeat message at one of our quarterly meetings. But after the presentation, I got to see how he treated the employees around him, and he was a very different person. People watch you closely, you live in a fish bowl, and if what they see is too different from what you say, they'll peg you as a phony and be less likely to believe what you're saying.

Speaking Hard Truths

The degree to which people perceive you as an effective or strong leader is also in direct proportion to how willing you are to talk straight to them. Although diplomacy and message spin have their place, leading in situations often means acknowledging out loud what many in the people know to be true but may be reluctant to address.

Your employees are often way ahead of you. If some discontent or worry about the company's future has trickled up to the executive suite, you can bet the concern is already rampant down on the front lines. Leaders gain a lot of credibility with the troops through honest acknowledgment of what's really happening in the organization, recognizing that the people they're speaking to are adults and can handle the truth.

If you're not prepared to speak honestly about a situation, you should hold off until you are ready. Too many leaders stand up and say everything is fine when it isn't, and their credibility takes a big hit. You send a message that all you're going to hear from the top of the organization is the company line, and it's hard to lead from that position.

Where leaders often get into trouble is when they haven't bought into their own messages and their body language betrays their insecurity or uncertainty. Most business people are not great actors, and we all know there is a disconnect between the verbal and nonverbal. People tend to believe the nonverbal almost every time. Leaders need to get clear on how they feel emotionally about their messages, then opt for as much honesty as possible, without spilling trade secrets or creating unnecessary gloom or panic.

Change is never comfortable, but when managers get up in front of their people and use all these wonderful platitudes about how change is so good for them, most people in the audience are only thinking about how painful it is. You need to first acknowledge how tough change is and talk about the struggles the group may face in the transition before those better times arrive.

Motivating Beyond the Numbers

As a leader of others, influencing minds also requires appealing to something beyond meeting your goals. Those skilled at using the corporate bully pulpit understand that most people need a reason to come to work each day beyond the bliss of doing your job well or cranking out more widgets than last week.

Corporate leaders often shy away from commitment to a higher cause because of the emotional nature of such appeals. But people yearn for that because most want to serve something larger than ourselves. There's nothing more demoralizing than a leader who can't clearly communicate why you are doing what you are doing. When you believe what you are doing is meaningful, you can cut through exhaustion, irritation, or fear and keep going. A constant emphasis only on financial goals doesn't motivate most people to work harder or longer hours, particularly when they see senior executives get fired and receive severance packages that are more than they'll earn in a lifetime.

Many managers miss opportunities to appeal to employees' hearts and minds in their communication strategies. It's important for leaders to talk about how your products or services can improve the quality of people's lives, rather than trying to inspire and lead the troops solely by focusing on exceeding last year's profit margins. It's not always easy to do without feeling corny or like you're pandering, but it's important for people to believe they're working for something other than the bottom line.

Feeling Their Pain

One's ability to lead is also a function of an audience's belief that you have taken time to view the world through their eyes. The leader who doesn't understand and empathize with audiences is going to be leading alone. It doesn't mean you have to agree with their points of view, only that you acknowledge and understand them. The idea is to find some common ground. Once they feel that you understand their pain, it's easier for you to lead them where you want to go.

Establishing an emotional connection with your team doesn't mean you have to cry on their behalf or otherwise emote. An indication that you are human and have walked in their shoes is often enough.

It's astounding how, when people feel cared about, respected, and understood, whether you're talking to them in one-on-one settings or in large groups, they become far more willing to follow and be loyal to you. Stephen Covey has a principle that I use regularly: "First, seek to understand before you seek to be understood."

Leadership is above all about creating relationships, a fact all leaders would be wise to remember in their communication approaches. Leadership is not simply sharing information, but sharing a relationship.

There are traps you should learn to avoid, however. Leaders in particular need to tread carefully when portraying themselves as sharing the same struggles as the rank and file, or when using anecdotes designed to humanize them. I heard one CEO try to share the pain about bad traffic by mentioning how his limo got caught in gridlock on the way to the meeting. While a certain amount of making yourself human can establish commonality with the audience, you have to be careful not to do it in ways that set yourself apart from them. Ringing a false note of empathy can be worse than making no noise at all.

It's Not All About Me

Some make it to the middle of the ladder of success and forget about the "little people" who helped them get there. They are suddenly too good to associate with the employees. Remember not to be a snob. You need to still be willing to do the dirty work or the lowly jobs.

Separation

Once you get promoted, you can't be the close friend you once were with your peers. You can still be friendly, but not fishing buddies. If you become too close, you show partiality such that others feel that they have no chance to advance because of your special friend. Also, don't let your friend get all the choice job assignments. Assign work to the best person to do the task.

It's All About Performance

The hard part of being a manager with your peers is communicating bad performance. Communicating good performance is easy. Communicating bad performance is why you get the big bucks as a manager. You need to communicate performance, whether it is good or bad, in a place that is familiar and business-like for the recipient. This is particularly true when communicating bad performance. It is a good idea to script out at least a bullet point list of performance issues to communicate to the employee. If it turns emotional, keeps your wits about you and keep your voice calm. If you expect a confrontation, consider having a human resources representative present. It's amazing how the mere presence of human resources calms people down.

Oxymoron No More: The Self-Aware Leader

People are more apt to trust, believe in, and like leaders who don't try to create an air of infallibility. Authentic leaders have a greater awareness of their strengths and weaknesses, so we know we can rely on them in their strong areas and that they won't put us at risk by stepping into something they aren't prepared to deal with. You trust and follow them because they don't put up a facade.

Acknowledging that you don't have all the answers and that effective leadership is a shared enterprise also can make people more eager to board your bandwagon. John F. Kennedy's "ask not" inaugural speech still stands as the shining example of this approach. Instead of always feeding them the answer, make them come up with the answer. Ask them, "What are you going to do to fix the problem?"

Leaders need you to help their subordinates understand that they can make a difference. The best leaders admit they can't do it alone and that they need to tap the talents of others around them in pursuit of a shared vision.

Authenticity also is about being secure enough to be natural and comfortable, even if it's not going to win you any acting awards. Being who you are always goes a long way in communication with subordinates. The result may not land you on the cover of *Time* magazine, but it's more likely to help establish your reputation as a leader people want to follow.

Do You Really Want to Be a Leader?

Leading can be awfully lonely and terribly frustrating. I haven't always been cynical. As president of the Student Engineer's Council at Texas A&M University in 1981, I thought leaders led a life ensconced in thoughts about the distant future. Everyone dearly loved to be around them. Leaders were surrounded by people who were contented, smiling, and appreciative, seeking a willing mentor. Leaders led a life of thinking, studying, relaxed research of the latest trends, unchallenged respect, and endless harmony. No conflicts, no arguments—after all, you are a leader and people listen to you.

Alas, reality set in. It is truly amazing what time, a few gray hairs, a lot less hair, and a passel of useless theories can do to a person. Today, there are more people who want to become leaders. When asked why, they really don't know, except that they think they will be paid more money. Precious few know that you get more abuse as a leader. A good leader takes a little more of the blame, a little less of the credit, and does a lot more of the work.

To those interested in becoming a leader, I would ask them five questions:

1. Do other people's failures annoy me or challenge me?
2. Am I using people in my interaction or am I cultivating and mentoring them?
3. Do I direct people or do I develop them?
4. Do I criticize or do I encourage?
5. Do I shun controversy or judiciously pick my battles?

It is not that leaders are not needed—good leaders are definitely needed. The problem is that it is a lonelier job than it once was. It is not like the times of people like General George Patton or Dwight D. Eisenhower. Part of this is to be expected. You can't look into the sun too long and not start to get arrogant. You need a good friend who will keep you humble so you will not be tempted to abuse the privilege of leadership. The old saw is true: power corrupts and absolute power corrupts absolutely.

When you choose to become a leader, you surrender yourself to others. You have to have a listening ear, a strong will to discipline, be able sculpt souls with a thundering velvet hand, and to undergo severe scrutiny, all without taking it personally. If you are very sensitive and have a thin skin, it will be tough to overcome.

There are frustrations. You will tell people who ask for advice what to do, and they will not follow it. You will try to shape people and they will resist, even though it is for their own good. You will try to show people where the land mines are and they will step on them anyway. You will see people fail, and fail, and fail.

Are you cut out for this tiring, thankless task? You must carefully examine yourself and decide if it is worth the sacrifice.

Continuing on, when communicating with subordinates, the following principles also apply.

Consistency. Consistency revolves around expectancy. That is, the expectation that a particular cause always generates the same effect. Further, whatever you do for one person you do for everyone under similar repeated circumstances.

Justice and Fairness. Similar to consistency, no one gets treated differently, rather all are treated justly and fairly. No one gets a break because of who they are, where they are located, and who they know.

Truthfulness. You must be truthful or you risk undermining your whole character. For example, when conducting incident investigations, don't attempt to hide the facts or blame others unless it is excruciatingly clear the incident was a result of a deliberate act. The same is true when reporting information to superiors. Character is followed closely by integrity. Without being truthful, you cannot have character or integrity. It has been said that the true measure of a person's integrity is what they do when no one is looking. Integrity is not believing that morals are pictures on walls or scruples are Russian money.

Leadership by Example. An SH&E manager must talk the talk *and* walk the walk. Employees look at you and judge your behavior according to their perception of right and wrong. You as SH&E manager must be performing far enough above the rules that it is clear that you are obeying the rules.

This is a difficult job in today's corporate environment. Communicating with your subordinates is crucial to your ultimate success or failure. A few false moves and you can lose your credibility and respect. Once that occurs, you may never be able to recover.

Check Your Culture IQ

Corporate culture encompasses the unwritten rules and informal norms that shape the way things are done in a company. Its unspoken message is: Fit in or foul out.

What's gone wrong in the following scenario? A manager is knowledgeable, hard working, motivated, qualified, and headed for failure. The staff is not performing up to expectations, goals are not being met, stakeholders are unhappy, and the harder the manager tries to fix things, the worse they get.

Complicating an analysis of the problem is the fact that the individual has been a success in previous positions, but is now viewed as a loser and probably will be forced out of the current assignment. The manager is either denying or defying the company's culture. Possibly without realizing it, he or she is fighting the values, beliefs, and assumptions shared by members of the organization.

Culture will either make or break your business plan, and will make or break you as a leader. If a person has the right style and is in the right culture at the right time, he or she is going to look like a hero. But put that same individual in the wrong culture and he or she is doomed. Individuals in companies don't pay enough attention to the culture.

These individuals don't "map the territory," which is fundamental. Look around and ask questions such as these:

- What actions and behaviors are rewarded in this organization?
- To whom do employees listen?
- What aspects of work do employees joke about, and what aspects are not considered appropriate to joke about?
- What obstacles does the culture create? What opportunities does it provide?

Learning The Three Rs

Mapping constitutes the first of the three R's of thriving in corporate culture:

1. Reading the culture,
2. Reading yourself and how you fit (or don't fit) in, and
3. Reacting realistically and appropriately.

There's no room for complacency in the process, so don't assume that you can take the culture for granted. Cultural forces and pressures can change while an individual stays in place. Reorganizations, mergers, and changes in top management shape a company's fortunes, and new product categories and revised strategies change, which can affect the way things are done in a company.

Meetings provide a window into the culture of an organization. Do the meetings have clear agendas? Do people feel free to speak up? Do a few individuals dominate the meetings? What issues get the most attention? Are decisions made at the meetings? Are follow-up activities assigned? What is the mood during and after the meeting?

Company stories, the teaching parables of corporate life, are another sure, fast way to identify organizational culture. They provide guidelines, lessons and advice by setting an example and posting a warning.

If someone asked you for three stories that exemplified the character of your company, what tales would you tell? The more revealing and on-target your stories are, the higher your cultural IQ, and the better your prospects are for doing the right thing at the right time.

As one example, here is a signature story that I heard that epitomizes the culture of a grocery chain. It centers on a senior executive who was visiting one of his stores when an angry customer confronted him with a plastic bag filled with rotten shrimp that his wife had bought for a dinner party.

The executive promptly gave him his money back, as well as twice as many shrimp. As the man was walking out the door, the executive said to him, "By the way, the next time your wife buys shrimp, have her come to our store." The executive pointed to the label on the plastic bag. The man had come to the wrong store to complain.

The story circulated quickly throughout the organization, sending a powerful message from senior management that the company goes all-out to serve both existing and potential customers. No management rhetoric or sloganeering could match the impact of that story, which became embedded in the grocery chain's culture.

When you hear positive stories like this about the culture of your organization, ask yourself how you would react in such situations. In the case of negative stories, how would you have behaved? Or, if you've faced a similar SH&E-related situation, how have you handled it?

This is one step in the direction of reading yourself. Another involves examining your previous successes and failures and the settings in which they occurred. When doing such self-evaluations, make sure you ask for input from others.

You may already have help in this process if your company is using a 360-degree performance review to collect feedback from the full range of people who deal with you. To maintain feedback, a buddy system is a powerful resource.

Find and stay in close touch with people you trust and know well, and who get to see you in action in different situations. They should not be individuals with whom you compete, but they could be mentors and coaches. When you feel comfortable with such individuals and believe they have your best interests at heart, you can confidently accept their feedback.

Adjusting to the Culture

Obviously, fitting in doesn't mean surrendering your identity. It does mean adjusting what you say and do in terms of the dominant culture. The operative trait is flexibility: the capacity to adjust to reality, working to change what can be changed, and learning to accept what can't be.

When interacting with others, the people who are most successful manage and communicate in an appropriate style. They socialize smoothly with others in the company. That includes talking about non-work topics, since small talk provides the social cement that helps bind people together.

Some firms are more genteel and diplomatic in the way they communicate. In others, you can be an elephant in a china closet and, if you're right, you're right, period. The difference between the two extremes calls for a difference in behavior.

Another key cultural difference centers on the ability to work in teams and to feel comfortable with consensus-building. However, though you have to play politics to move up the ladder, you won't succeed if you compromise too much and don't take a stand on certain principles. You need to find the right balance, to be able to judge when it's critical to take a firm stand, and when playing ball is the best approach.

Raising your culture IQ is a do-it-yourself challenge that shouldn't be taken for granted. Most companies won't tell you about their culture, so it's up to you to find out.

Cultural Signposts

Every organization has cultural signposts that indicate the dominant or preferred styles of operating. They tell you what behavior is appropriate, and what is reinforced and rewarded. Taken together, these signposts identify a company's cultural tilt.

The following questions may help you profile your organization and evaluate how well you fit in:

- Does your company have a formal or informal style of interacting and working?
- Is it oriented toward teams or individuals?
- Does management have open or closed communications with staff?
- Is there a silo mentality or is the operation seamless?

- Is your firm customer-oriented or organization-directed?
- Does management foster a confrontational or avoidance style?
- Is the firm politically or issues-oriented? Does the substance or the politics of an issue dominate?
- Is the culture freewheeling or controlled?
- Is the company oriented toward change or toward maintaining the status quo?
- Does management focus on goals or tasks?

Eight Power Points for Presenting More Confidently

No matter where you are on the corporate ladder, confident presentation skills will get you noticed, remembered and promoted. By developing good, solid, speaking skills, you increase your value to your company. You will also increase your career options and improve your professional and personal life. Not sure how to get started on these lofty goals? Begin by practicing these eight tips to help you present yourself more confidently.

1. Seize the opportunity

A key building block for developing confidence as a speaker is to speak, and speak often. Seize every opportunity you can, personally and professionally, to speak in public. If someone invites you to "stand up and say a few words," or a co-worker asks you to make a presentation, jump at the chance. In fact, don't wait to be asked, volunteer!

2. Use the "as if" principle

If you want to be a persuasive presenter, start acting as if you are. Dress, speak, and behave as a confident speaker would. Assert your knowledge and expertise by speaking up in meetings, contributing articles to company or trade publications, and positioning yourself as **the** presenter on a particular subject. When you think and act as though something is true, you help make it happen.

3. Realize you are the expert

If someone asks you to speak or give a presentation, there's a reason—namely, that people perceive you as an authority on a subject and they want to hear what you have to say. That should give you some self-assurance. Trust yourself as a presenter and you'll project confidence.

4. Meet your audience before you present

A good way to build your confidence (and instill a great first impression) is to arrive early and, as guests enter the room, introduce yourself, shake hands, smile, and look them in the eyes. You will be surprised how this exercise rids you of nervousness. It also sets the tone for a relaxed, natural delivery, making your presentation seem more like an extended conversation among friendly people.

5. Visualize your success

Like a World Cup skier preparing for a downhill run, before any presentation, mentally walk your body and emotions through your talk. See yourself speaking with confidence and poise; hear yourself speaking with eloquence; feel your energy as you stand before an enthusiastic audience. Be the presentation. Your body will respond to the pictures you hold in your mind. Then, when it is time to perform the real presentation, your thoughts and emotions are in control because you know you've been there before.

6. Make anxiety your ally

Many of us get a pounding heart, buckling knees, sweaty palms, a dry mouth, and butterflies as pre-speech symptoms. These are nature's way of preparing you for action. The key to conquering anxiety is not to abolish it, but to learn to use it effectively. Those jittery feelings are the very tools you need to make a dynamic presentation. They increase your energy, heighten your awareness, and sharpen your intellect. Rather than squandering this energy in fear, use this natural, physiological reaction to think faster and to talk more fluently, with greater intensity.

7. Rehearse, rehearse, and rehearse some more

As the nineteenth-century essayist William Hazlitt wrote, "We never do anything well till we cease to think about the manner of doing it." Rehearsal familiarizes your mind and body with the mechanics of presenting. Practice frees you to focus on the message, not the manner of delivering it. This way, during the real event, you are less self-conscious and more audience-conscious. So be sure to rehearse your presentation out loud, even in front of a dress rehearsal audience if you can. Use your spouse or a close friend as a sounding-board audience. You need someone who will be completely honest with you regarding ways to improve. This will keep you humble. Getting a third disinterested party will help you ferret out what works and what doesn't.

8. Realize that your audience wants you to succeed

Listeners respond to you based on their own self-interests, which means that instead of critiquing your speaking abilities, they're concentrating on your message and how it can benefit them. So it helps to remember that you and your audience are on the same team, advancing together toward a solution. When you give a winning presentation, they win, too.

Compelling Communications to Upper Management

One way to demonstrate your leadership, business skills, and professionalism is when you communicate with upper management. The dilemma is that in SH&E, we deal with a lot of data and we struggle with how to communicate it to upper management. The problem is that we lose our message by not condensing all that data into usable and understandable information.

Here is a typical scenario. You are on your boss's calendar and as you enter the office, your boss says, "Make it quick, I have a conference call in 10 minutes. What do you want?" If you are not prepared to present your case concisely and understandably, your message is lost and often your initiative denied. What we often fail to realize is that when we catch our boss, we often have about 90 seconds to either win or lose, or go back and do more research. This is our chance to make a good impression. If we miss, we might never get the chance again.

We must be concise and relevant with our verbal and written word. Our memos must be short but clear. Our reports must convey relevant information, not a diatribe of data to be sorted through. Our presentations must be the lightning strike that provides an absolutely clear picture of the problem. Doesn't this sound easy? Well, it isn't for many of us.

What I have included here are lessons I have learned over 25 years of trial and error regarding the efficient but sometimes ineffective use of presentations, what we say, and what we write for CEOs. Our goal is to get the boss to say, "Oh yes, now I see what you want. By all means press forward."

We all struggle with this as we live the in the lightning-fast business lane. This information will help you craft your ideas in a fashion that will get the attention of your bosses in a way that they view you as the person who knows the issues and solutions. You will learn to do all of this without wasting the boss's time.

Bosses usually want to know a few things about what we are proposing, such as:

- How do we compare to our competition?
- How does SH&E contribute to the company's financial health?
- What is the ROI?
- What needs to be done and why?
- What are the next steps?

When we approach our boss we need to consider these questions. We need to frame our proposals around these questions in terms that are simple and relevant to the reader, not just to SH&E professionals.

The Written Word

There is a temptation to provide all the data to our bosses so they can see what we see. We think that surely they will agree. Sadly, they don't understand all the underpinnings of the data. We need to take all the data and shrink it into compelling information. This information should be succinct, clear, and obvious. Here are some strategies:

Be clear. Don't set up the story with a lot of history. Cut to the chase right from the start. This means you have to think the problem through and know where you are going. If you wander around it will raise the question, "Where are you going with this?"

Be brief. Most of our emails are too long. We need to think about keeping them to one page. Emails should be about three-quarters to half a viewing page. Try to keep you message to one page if at all possible.

Use short sentences. Long sentences lose their effectiveness quickly. Use sentences of fifteen words or less to communicate your message. This means you have to work at it. Write it down, walk away, revisit it, and revise it. Once it is ready you are now prepared.

Avoid the fog. We are tempted to use all the SH&E jargon such as TRIR, DAFW, SPCC, and Tier II. This often complicates our communication. Define the terms that make it relevant. As I will demonstrate later, present this information in light of regulatory, company goal, and peer company measures. This makes your metrics real and relevant to the reader, your boss, your peers, and other upper management.

Write from their point of view. Using the questions I have shown above, frame your memos and reports with these items in mind and address them in some form or fashion.

Seize the reader's attention. If you can, link to a significant event that the reader can relate to. Use that event to start it off, whether it is internal or external to the company. For example, "In light of the recent fires caused by electrical failures near our facilities, which have cost an estimated $500,000 in damages, we need to address our electrical distribution system." Or, "Lightning strikes have cost the company over $1M in lost product over the recent months. As a result, we need to address our high-exposure facilities and protect them."

Have the data available. As stated earlier in this chapter, we must have the data available, perhaps in attachments, but not in the body of the memo or report. This helps communicate that you have done your homework.

If you do this, your memos and reports will be well received and actually read. And when your boss is done, he or she will likely have a clearer understanding of the problems you are facing and the solutions you are proposing.

Your Verbal Presentation

When we present to our bosses we have to overcome some common barriers. These barriers prevent us from accurately communicating. Some of the barriers include:

Perception differences. What we mean by certain terms are not what they understand them to be. Their view of the world is usually external or upward in direction. Our world is more internal. We need to frame our discussions around their perceptions, not ours.

Overload. We live in a world of stimulus overload. We need to understand that overload and be sensitive to those issues. If your boss is in the middle of something, you may want to wait for a better time and rethink your message to be right to the point.

Speed of listening. Some bosses listen more slowly, other listen more quickly. The ones who listen slowly will stop you and ask you to go back and explain things. The ones who listen more quickly will often jump ahead of you or ask you to speed it up so they can respond. They have often already formed an opinion or answer. Be attuned to these characteristics and proceed accordingly.

Timing. Be sensitive to when you approach your boss. Depending on your boss's role, such things as the annual report, 10K and 10Q reports, and quarterly board meetings have very restrictive timelines. Avoid the due dates for these items. Find out the best time to approach your boss and you will have a more positive reception for your proposal.

Trigger Words. Know your boss's trigger words, like "increase the budget," "more personnel," "out of compliance," "SH&E risk," and "risk management exposure"—the list goes on. If you know these trigger words you can lace your verbal discussion with them to get your boss to key in on your needs.

Semantics. In SH&E we often get caught up in semantics. What we meant and what they understood become miles apart. We need to couch our arguments in terms relevant to the listener, just like we do with our written words.

Not listening for attitudes. When we fail to listen for attitudes we can jeopardize our message. If we pick up on the attitudes, we can adjust our message to strike our target and positively communicate our message. Failing to listen for attitudes can cause our message to fall on deaf ears.

It is known that, when we communicate, seven percent is communicated by words, about 38 percent by intonation, and 55 percent is non-verbal. Keeping attuned to the 93 percent rather than just our verbal delivery can help us succeed in verbal communication.

In the end, we want to avoid getting the preemptive strike from our boss when he or she says in the middle of our discussion, "Deliver the baby but spare me the labor."

Listening. If we fail to listen when our boss talks and only focus on our message, we are doomed to lose. We must be motivated to listen. To do this we must not interrupt. This is hard for many of us, but we must let our boss talk. After all, communication is a two-way street. This also means we must fight off distractions of anything in the background and other conversations. The best way is to give as much eye contact as possible. Listening and adjusting our message or re-framing our message can mean the success or failure of our request. Listen and listen hard.

Presentations

What I am specifically referring to is Microsoft PowerPoint. We have fallen into a trap with PowerPoint. Our slides look the way they do because PowerPoint has left us in a place where we don't think about how we present information. What I will show you here is based on trial and error and adjusting your presentation of information to the culture of different companies to make the information relevant to the reader.

Often the data changes only a little, but I have sorted the data, eliminated a lot of it, and created information. This information makes my presentations compelling and relevant. In fact, I would go so far as to say that many of us have failed in communicating, myself included, because we have failed to present our arguments in a compelling visual fashion. On the other hand, I have literally seen my credibility rise from the unknown or even questionable to the point that management comes to me because I was so well connected to the field. All this was due to my ability to present the data in a way that was relevant and compelling. The bottom line is to present a case that the desired decision by upper management becomes as obvious to them as it already is to me.

I have used several key references that I recommend to you as required reading to accomplish this goal. They include three works by Edward R. Tufte, *Envisioning Information*, *Visual Explanations*, and *The Visual Display of Quantitative Information*, and two by John W. Tukey, *Exploratory Data Analysis* and "Summarization: Smoothing; Supplemented Views" in *Interpreting Multivariate Data* (with P.A. Tukey).

- Tufte, Edward R. *Envisioning Information*. Cheshire, Conn.: Graphics Press, 1990.
- Tufte, Edward R. *Visual Explanations*. Cheshire, Conn.: Graphics Press, 1997.
- Tufte, Edward R. *The Visual Display of Quantitative Information*. Cheshire, Conn.: Graphics Press, 2001.

- Tukey, John W. Exploratory Data Analysis. Reading, MA: Addison-Wesley Pub. Co. 1977.
- Tukey, John W. and P.A. Tukey. "Summarization: Smoothing; Supplemented Views" in *Interpreting Multivariate Data*, edited by Vic Barnett. New York: Wiley, 1981.

The key reference is Tufte's *The Visual Display of Information*, which can be purchased online. This book can also be purchased with a PowerPoint tutorial. Reading the book and studying the PowerPoint examples will go a long way to improving your use of PowerPoint. I know it certainly has for me.

Here are a few rules:

- Don't blindly rely on the automatic graphic formatting provided by Excel or PowerPoint! Look at what you are displaying and choose the form that best illustrates your point.
- Strive to make large data sets coherent. This is particularly true when we are looking at accident causes that have multiple small percentages of certain accidents.
- Encourage the eye to compare different data. Make the data relevant by comparing to similar data sets over years or similar projects.
- Representations of numbers should be directly proportional to their numerical quantities. The numbers should accurately reflect the numerical value in a pie chart or a bar graph.
- Use clear, detailed, and thorough labeling. Don't make the reviewer interpolate or wonder what the label means.
- Display the variation of data, not a variation of design. The variation must match the trends and provide additional data that explains more detail that the other chart.
- Maximize the data to ink ratio–put most of the ink to work telling about the data! Invest in the picture, not the number or table.

- When possible, use horizontal graphics: 50 percent wider than tall is usually best. This allows the reviewer to see the span the data includes, like months or years.

Tufte has a few remarks about information presentation, as follows:

- Visual reasoning occurs more effectively when relevant information is shown adjacent in the space within our eye-span.
- This is especially true for statistical data, where the fundamental analytical act is to make comparisons.
- The key point: "compared to what?"

Before and After PowerPoint Presentations

Here are a few examples of using these principles that illustrate how they can improve your PowerPoint in a way that will dazzle your boss and make him or her understand just what you are presenting.

Before. Figure 11 displays how PowerPoint usually places our data, in the form of a bar chart. In this example we are comparing the Worker's Compensation experience modifier of six business units over a seven-year period. We are looking at trends and trying to assess the relative performance of each business unit. As you can see, even for the trained observer, this is difficult at best using the bar chart from PowerPoint. So, can you tell how our business unit's performance stacks up? It is not clear at all.

Figure 11. Before–Employee Worker's Compensation experience modifier

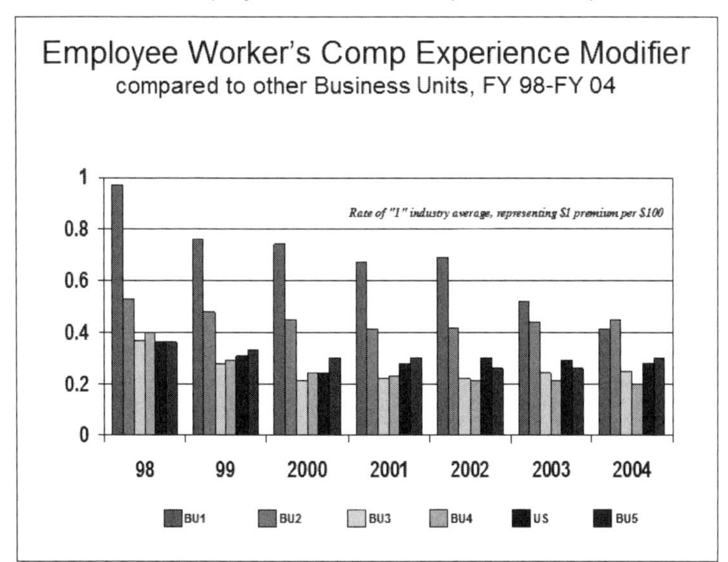

We've all seen these charts and have difficulty making heads or tails of them, and we know what the numbers mean. Try convincing upper management with this chart to take action or to demonstrate success, and they won't know what to think.

After. Figure 12 shows a PowerPoint chart where we have chosen the format in a way that clearly communicates our relative trend and success. You can clearly see the difference in clarity of presentation. It is now clear that our work comp EMR (US) is second among all other business units (BU1–5) presented and dropping to where we will soon be in the number one position. In Figure 11, the presentation is unclear even to the expert reviewer. In Figure 12, the trends jump out even to the uninitiated reviewer. This is what you want upper management to do. You want them to say, "Aha, now I get it." And you haven't even said a word. It is true, a picture (a good one, anyway) is worth a thousand words.

Figure 12. After–Employee Worker's Compensation experience modifier

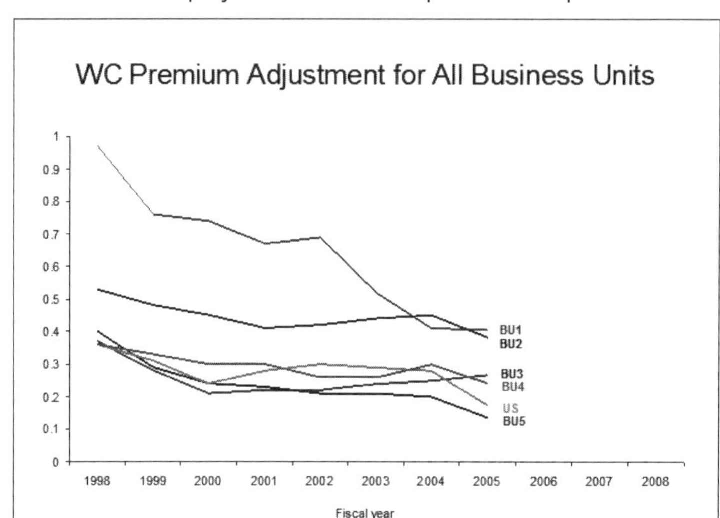

Before. Figure 13 displays another one that we deal with on a regular basis (perhaps monthly, but at least annually). This chart is the perennial chart of incidents by type. We've all struggled with how to deal with this, often having difficulty explaining the chart ourselves. This is the problem of displaying data rather than information. We feel compelled to illustrate every little incident we've ever had in the last year, significant or not. The result is what you see below, a chart with lines everywhere. No trends appear and most of it seems to be in the noise.

Figure 13. Before–Worker's Compensation injury type

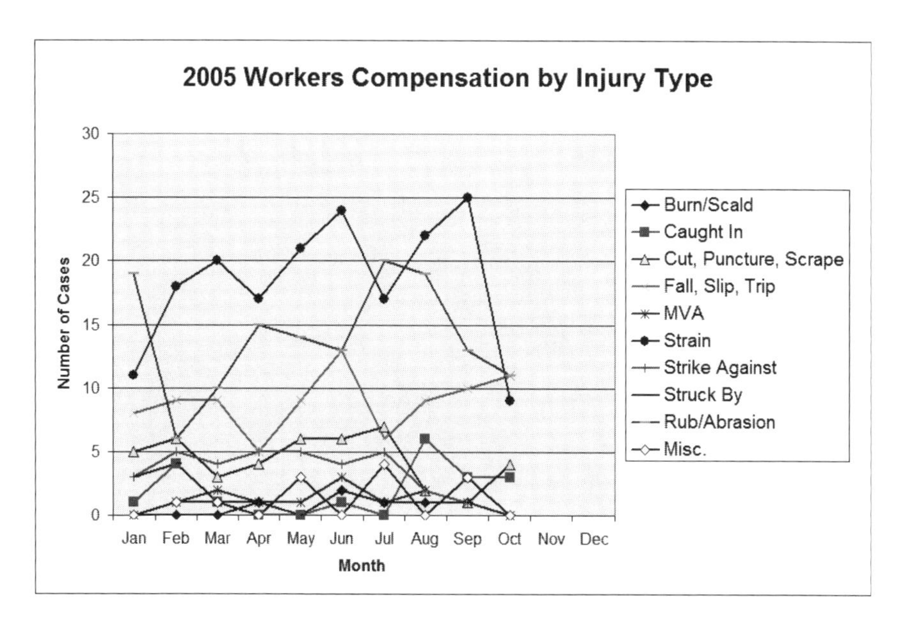

After. As you can see, in Figure 14 the data has been reduced to information. We no longer see every individual item. We see the important ones. We have eliminated the data and replaced it with information. This is what companies pay us to do. They pay us to look at the data intelligently and present information. This chart is compellingly simple and the significant trends obvious. This will keep upper management's attention and get them coming back for more. Your charts are enlightening, not baffling.

Figure 14. After–Worker's Compensation injury type

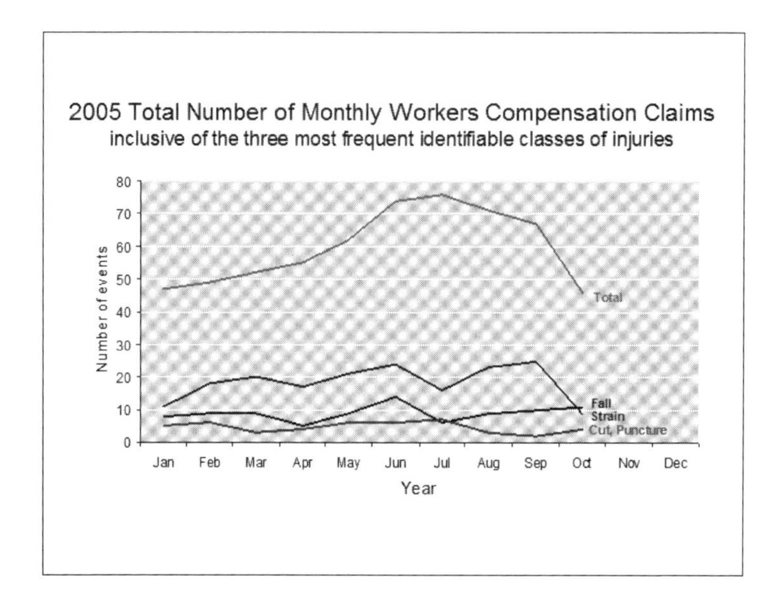

A few more examples are presented below that I have personally used. These are based on the company's information needs and cultural desire on how they review the information presented and what they want to see.

In Figure 15, we see a coupled design of accumulative incidents starting in January and progressing to December. The company management revolved around annual results, so I compared the data from 2005 to that accumulated in 2006. Further, you can see the pie chart of incident types for both 2005 and 2006 on the right. The notes allow you to add a few salient points to identify the meaning of the information trends.

Figure 15. 2005 vs. 2006 YTD cumulative incidents

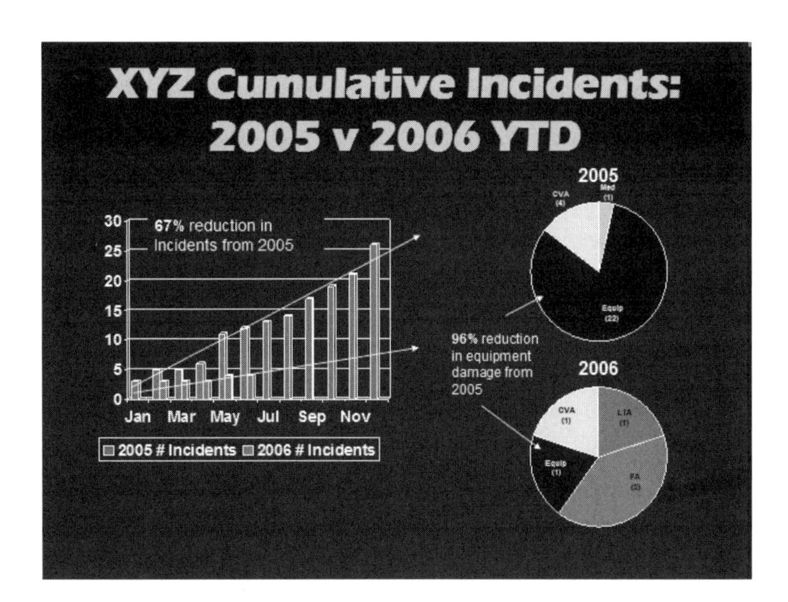

In Figure 16, we see a another accumulative chart, but this one for spills starting in January and progressing through June and then to September. Here we are showing the same kind of annual performance from 2005 to 2006 and accumulated from January to June, which is when the presentation was made in 2006. We can comment on the trends and potential trends for year ending if the current performance is sustained.

To make a point clear and distinct, we can separate severity (spill cost, Figure 17) and frequency (number of spills, Figure 18). Again, this is a progression from January to July and ultimately to December. Here we can see the increase in the costs from May to August and the steady increase in frequency. We can identify that between May and June we had the biggest increase in cost (severity) and number (frequency).

Figure 16. 2005 vs. 2006 YTD cumulative cost of spills

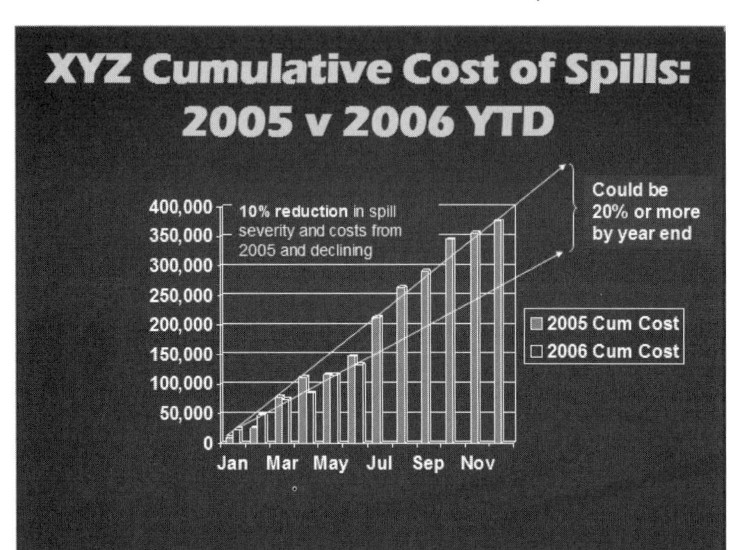

Figure 17. Spill cost YTD

Figure 18. Spill frequency YTD

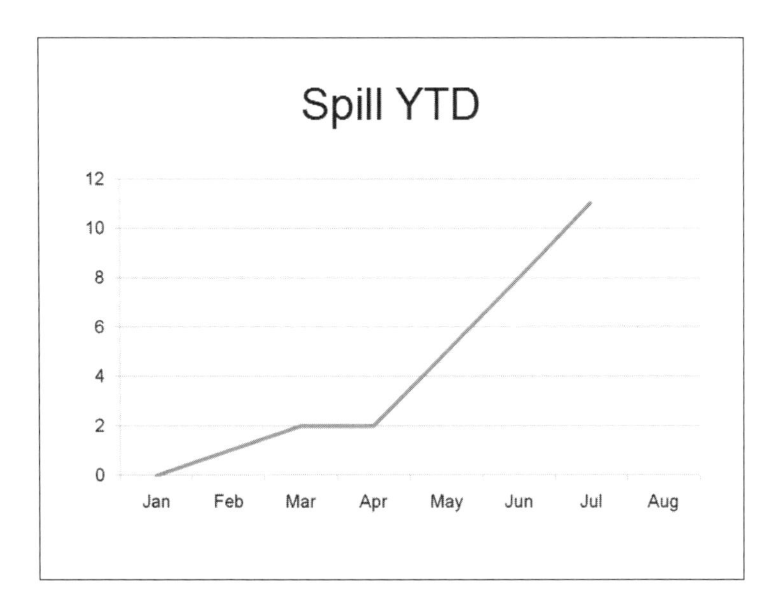

In Figure 19. we see a different view of the information in the form of rolling average. Some companies prefer this form because it more accurately reflects sustained performance than an annual accumulation. There are several reference points for relevance. First is the red, yellow, and green illustrating the company goals. Second is the industry average based on the most current information available. Third is the relative performance by business unit. It is obvious to see where the company appears compared to the industry and how each business unit compare to another and where performance issues are present.

Figure 19. Total recordable incident rate (TRIR)

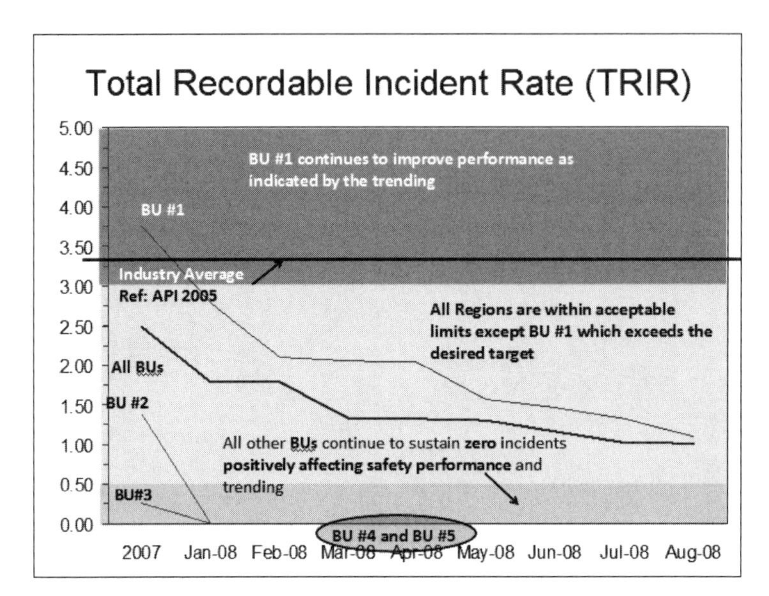

These are just a few examples to graphically illustrate quantitative data in the form of information. It is obvious to the reviewer and the pictures speak for themselves.

Although these techniques are powerful, there are some downsides to this approach. It takes a lot of time to create and assemble the data and create non-standard graphs that may not mesh with your work demands. You have to choose to say, "enough is enough." If you don't, you could get caught up in relentless tinkering and artistic judgment.

Smart Quotes Can Inspire the Creative Genius Within

Quotations are interesting things. They can really spice up a presentation, article, or speech. It has been said that the evolution of a quote goes like this: the first time it is quoted from the citation; the second time, it is quoted from a known source; and the third time it becomes yours. English scholar Andrew Lang once observed, "He uses statistics as a drunken man uses lampposts, for support rather than illumination."

So it is with quotations. Most often they are used to introduce or support a premise made in a presentation, sales pitch to management, or advertisement for an SH&E initiative. Unfortunately, if you are using them solely in that context, you could be overlooking the inherent value of the quotations themselves. In fact, quotations can be indispensable aids in developing presentations, not just as a means of support, but also as a source of illumination. They can be an invaluable brainstorming and authoring tool, and a vital part of the creative process.

Good Quotes Get the Creativity Flowing

To illustrate, years ago I was asked to make a short presentation at a three-day regional conference where the theme was "Success in the '90s."

To get the creative juices flowing, I pulled out a copy of my favorite book of quotations and searched the section on success quotes. I came across this quote by John Charles Salak. "Failures are divided into two classes: those who thought and never did, and those who did and never thought."

Starting with this single germ of an idea, I developed a presentation called, "Three Guaranteed Ways to FAIL in the '90s or any Other Decade."

Assuming the other speakers would discuss how to be successful in the '90s (an assumption that proved to be correct), I decided to take this more novel approach to set my presentation apart from the pack.

Since this program turned out to be the best-attended and highest-rated of the conference, I give much of the credit to Mr. Salak. His noteworthy and succinct observation saved me immeasurable hours of development time.

After that, even more ideas flowed from his basic premise. Using his quote as a starting point, I came up with three, rather than two, types of failures, and addressed each of them in the form of a sleep metaphor:

- **Dreamers**—they think, but never do. Those are the Walter Mittys in life who have all kinds of grandiose ideas and brainstorms, but never follow through due to procrastination, lack of self-confidence, or persistence.
- **Sleepwalkers**—they do, but never think. These are the Ralph Kramdens of the world, who run with an idea without thinking it through or without taking the time to plan and prepare.
- **Zombies**—they never think or do. I can't prove it, but I suspect a demographic survey would reveal that this segment of the population consists largely of people who watch too many TV infomercials. More specifically, these are the folks who simply wander through life without any specific goals, aim, or purpose.

Sometimes Your Message Will Write Itself

I then found other quotations on the subjects of procrastination, planning, goal-setting, and so forth. So, in my own way, I addressed many of the same issues the other speakers did, but from a different frame of reference. And it was this approach, initially fueled by one solitary quotation, that ultimately made my presentation unique.

Best of all, my presentation almost wrote itself. What would probably have taken many hours (if not days), took only a few hours, and my presentation, I am certain, was infinitely better because of it.

I'll close with this advice from English clergyman and writer, Charles Caleb Colton. "When in reading we meet with any maxim that may be of use, we should take it for our own, and make an immediate application of it, as we would of the advice of a friend whom we have purposely consulted."

So, come on, wake up and take a look at quotations in a new light. You'll find that Samuel Johnson was right. "Classical quotation is the parole of literary men all over the world."

Learning to Manage Like a Coach

A manager with good coaching skills is long remembered, greatly admired, and amply rewarded. Fortunately, coaching is a skill that can be learned. Your favorite teacher probably coached you to bring out your best performance as a student. Other skills, such as selling and riding a bike, are also acquired through instruction, practice, and hands-on coaching. In fact, that is how we become proficient at any skill or task.

As a manager, your goal is to help the people on your staff become better performers, so coaching is definitely part of your role. In developing your career, you should consider coaching from two perspectives:

1. Coaching to lead others: The results you obtain from other people correlate with your ability to coach.
2. Being coached to achieve self-improvement: Change is constant, and the actions you used effectively in the past may no longer produce successful results. Therefore, you must continually refine your personal skills. Accepting coaching affords you opportunities to grow and learn.

How Coaching Works

Coaching is a powerful tool that can lead to genuine performance improvement. It often involves an ongoing relationship that emphasizes listening, understanding, and bringing internal obstacles to the surface. Like any other skill or process, effective coaching must be learned and practiced. Here are some recommendations to help you improve your coaching ability:

Define objectives. The coaching relationship should be based on the needs, values, and goals of the person being coached. Develop team and individual objectives and break them down into step-by-step actions that can be measured periodically.

Communicate clear expectations. Clarify direction, goals, and accountability as they relate to specific performance expectations, and explain why these are desirable both for the individual and the organization. Have high expectations for yourself and for others, encourage risk taking, use failure as a learning experience, and always let employees know where they stand. Communicate openly and encourage mutual respect, individual responsibility, and participation in team activities.

Evaluate needs. Evaluate the person you are coaching to determine his or her knowledge of what to do, skillfulness at execution, willingness to perform, confidence level, and any barriers limiting performance.

Involve the individual or team. Involve them in the option creation and decision-making processes to foster accountability. If new actions fail, encourage the person or group to try something else. Link the organization's success with the individual's to gain support for even the toughest decisions.

Coach. Execute improvement methods in stages: Give advice, coach for skill-building, create challenge, invent better tools, and remove performance barriers.

Encourage peer coaching. Teammates all have a stake in and are accountable to the common goal. Peers can teach others to carry their own weight, innovate, and develop their skills.

Help team members develop the skills to cope with change, the discipline to persevere, the confidence to withstand uncertainty, and the courage to initiate and innovate.

Deal with emotional obstacles to maximum performance. If people get stuck, confused, or frightened about something, review what stops them and point out the ways that they hold themselves back.

Enable skills development. Provide training, tools, and the other resources people need to develop their skills.

Evaluate performance and give feedback. Evaluate team and individual performance based on agreed-upon expectations. Phrase feedback in terms of observed behaviors and point toward solutions instead of critiquing errors. Acknowledge successes and reinforce improved behavior and successful results. If necessary, repeat the cycle of creating, choosing, and acting on a new option.

Reward achievement and be creative in customizing the rewards. Keeping the coaching interaction positive will help the relationship last.

Lead by example. If you want to cultivate empowerment, accountability and responsibility in others, you must demonstrate those behaviors yourself. Develop and practice your coaching skills. To achieve positive change, you must create it, and coaching is a powerful technique to enable change.

Coaching your employees will provide tremendous dividends, you will be a better manager, and your employees will grow without resenting you pushing their envelope of comfort. Why not start now?

Personnel Development and Delegation

Personnel Development

Leaders understand people and their various needs in the workplace. Leaders understand that people and their needs are different. Leaders are able to identify what people want from a job, a position, or a career. They are able to map out short- and long-term goals for their people, nurture a commitment from the group, and lead to success by example.

Skills needed for personnel development are:

Developing commitment. Leaders make a direct emotional connection with fellow professionals that goes far beyond the usual boss-subordinate relationship. Leaders involve others, seek advice, ask for information, solicit solutions to problems, and provide frequent positive feedback. They make people who work for them create solutions and take responsibility for outcomes.

Encouraging empowerment. Leaders encourage fellow professionals to be self-reliant. They delegate to other professionals, and provide others with an environment for success. That is, they lower barriers and provide the tools and support necessary to succeed. They also allow an environment of failure without fear and an opportunity to take limited risks. If professionals fear failure (they may lose their job) they will never take chances.

Ensuring success. Many managers hire people who are less intelligent than they are, fearing that they make take their position. Leaders, on the other hand, try to hire people smarter than they are. If they succeed, leaders get credit for hiring the right people. Leaders are always looking for their replacement. They understand that in order to move up in the organization, they need to groom their replacement, maybe more than one, for contingency purposes.

Inspiring lofty accomplishments. Leaders give credit to individuals for their successes outside the group, but take responsibility for the individual failures outside the group, with no excuses. Leaders use accomplishable goals to build greater and more difficult goals and to encourage people to do more. Each time a goal is set, they make the next level a little more difficult, constantly building to surpass previous achievements. Leaders look for the quantum leaps rather than the linear approach to performance improvement. They are passionate, committed, and tenacious toward goal setting and accomplishment.

Modeling appropriate behavior. Leaders have earned respect because they symbolize the values and norms of the group. They lead by example. They talk the talk and walk the walk. SH&E managers make the rules that they must follow without exception. They must also exhibit knowledge in the field. They must be respected in their field, just as any other professional should be.

Focusing attention on important issues. Leaders have the ability to ferret through all the facts and discover the key issues and tough problems. Leaders recognize that only a limited number of goals can be pursued at one time, so they take care in choosing what to emphasize.

Connecting their group to the outside world. A leader serves as a link to the rest of the organization and the rest of the world. Leaders represent the group to the outside world, project the image of the group to the outside world, and relay information to and from the outside world. They get involved with university professors, researchers, and technologists in order to keep up with technical issues and keep in touch with their counterparts at other companies.

Teaching professionals the nature of leadership. Many professionals spend years in college learning about a specific discipline, but little, if anything, about leadership. I've never seen a college course on leadership. Professionals need to know that there are tools for effectively managing and effective tools for leading. Professionals succeed in their chosen field, get promoted as a result of this success, and find that they have few management or leadership skills. They are potentially doomed to failure, or at least mediocrity, without proper training.

Putting professionals in the proper environment to learn leadership. Leaders allow their professionals to practice leadership in a positive learning environment. Just as a safety professional learns that classroom safety is far different in the real world, so must a safety professional learn how to apply hands-on leadership skills.

Delegating

Delegating is one of the most important skills you'll ever master. Delegating doesn't mean passing off work, it means giving others authority, responsibility, and accountability. People who fail to delegate burn out early. People who fail to delegate do not effectively match the right people with the tasks for the most efficient outcomes. No one person can try to do everything without wasting valuable resources of time, money, and morale.

Steps to effective delegation include:

Trusting your staff. One approach I use with my superiors when I am questioned is, "If you don't trust the people working for you, get rid of them and hire people you do trust." The same is true with professionals. Give them the opportunity to try and succeed, or fail and learn if necessary. Let them know that making honest mistakes will not be punished. Let them know that you are providing a safe environment for calculated risk-taking and decision making.

Avoid seeking perfection. Engineers sometimes seek perfection at all costs. At some point engineers must stop engineering and start implementing. This implies some amount of imperfection; however, the imperfection may be within acceptable boundaries. Also, over-engineering implies a continual evolution: if left to their own ends, engineers will design and design and design, never implementing, but always improving, making the design better.

Giving effective job instructions. The workplace is busy. We almost expect people to read our minds. This approach to communication predestines failure. Explicitly communicate goals and performance standards at the beginning of a project, and provide frequent feedback to ensure progress.

Recognizing the talent and ability of others to complete projects. We need to understand that others approach problems differently from us. In many cases, there are numerous paths to the same solution. Allow and be flexible to this variability in problem solving.

Recognizing skills. Recognize that some professionals have much stronger technical skills than leadership skills, and allow these individuals to freely develop their technical expertise.

Following up on progress. Failure to follow up is probably the most grievous error in delegating. Provide employees with clear milestones that include dates and tangible products. Require a contingency plan with priorities if milestones are not met. Follow-up here is critical. If these plans are not met, consider redelegating the tasks.

Praising the efforts of your staff. A simple "thanks," taking employees out to lunch, a brief memo to superiors, a pat on the back, and so forth are just small examples of how to say "thank you for a job well done." Also, remember, praise in public, punish in private.

Avoiding reverse delegation. If an employee has accepted the work that was delegated, do not allow the employee to offer incomplete work. Provide guidance to successfully complete the work.

Don't make delegation an all-or-nothing proposition. Some employees need to build confidence a little at a time. Allow employees to do pieces that they feel comfortable with and can succeed in to build confidence. Keep building on this base until employees feel competent in the work assigned.

Supporting your employees. Support your employees with the knowledge, resources, authority, and responsibility that you would require yourself. Tell employees what you expect from them and what they can expect from you.

Delegating to the lowest possible level. This ensures that you are making use of everyone's talents and applying your time the most effectively.

Some people can lead, some can delegate. Few can do both.

Assessing Personalities for Delegation and Performance

One of the ways I have successfully assessed personalities for performing assigned task in the most efficient manner without having to do psychoanalysis of each employee is by using psycho-geometrics from *Psycho-geometrics: The Science Of Understanding People, And The Art Of Communicating With Them* by Susan Dellinger. I know it sounds weird, but it works. At least it has worked for me. So here it is.

Choose your top two most favorite geometric shapes from the items below.

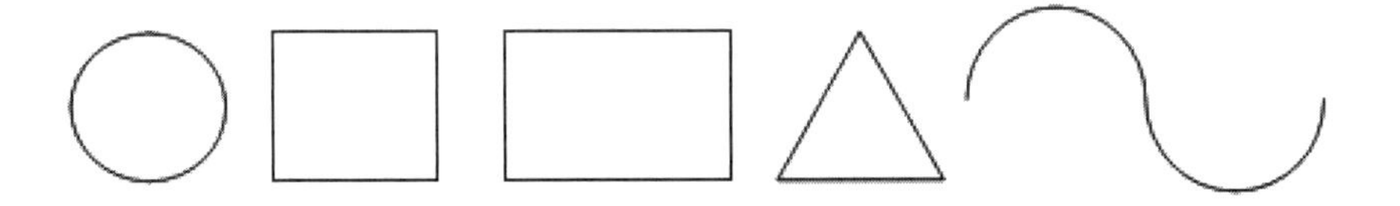

Hmm. So, what did you choose? Which figures are the most pleasing? Whichever they are provide interesting insights into your personality. These insights allow you to more appropriately allocate tasks to your staff, believe it or not. It's that simple.

If you selected a circle:

- A lover
- Right-brained
- Smooth

- Talkative
- Accommodating
- Adaptive
- Well-rounded
- Social, people-oriented
- Sequential
- A thinker
- Uniform
- A person who hates detail

A circle is someone who you would put on the planning of the company picnic, United Way drive, company skit, chair of the suggestion committee, and so on. They would like to meet all the people in the world. No one bores them. They like to be around people and like interacting with people. They are the people who enter a room filled with people and yell, "All right, a room full of people I haven't met yet!"

If you selected a square:

- A hard worker
- Left-brained
- Task-oriented
- Balanced
- Analytical
- Linear
- Logical
- Organized
- Sequential
- A thinker
- Uniform

- A person who attends to detail

A square is someone with distinct boundaries who doesn't like going beyond them. These people need instructions at every juncture and are typically introverted. They have met all the people in the world they ever want to meet. They are the people who enter a room filled with people and moan, "Oh no, a room full of people I haven't met yet. I wish I could just email them."

If you selected a rectangle:

- More flexible—wider base
- Left-brained
- Not as precise as a square
- A team player
- Fair
- Cooperative

A rectangle is somewhat more flexible than a square, but boundaries are still apparent and they are reticent to go beyond them.

If you selected a triangle:

- A hard worker
- Mid-left-brained
- Linear
- Power-oriented
- Moving up
- A person who looks at all the angles
- A person who likes to do several things at once
- A person who does things "ready, fire, aim"
- Task-oriented: get to the point

A triangle is someone who uses both sides of their brain and thinks that they can drive, talk on the cell phone, operate a PDA, and use their laptop in traffic and not get in an accident. They wish they had all the problems solved yesterday. They would like to rule the world to right all the wrongs in the world.

If you selected a sine wave:

- More flexible—wider base
- Right-brained
- Embraces change
- Lives by the phrase, "Fake it till you make it"
- Open and flexible
- Pictorial
- Leaves the details for last
- Non-directive
- Intuitive

A sine wave is your free thinker. This is the person you would ask to write the program for a new hazard where there was no previous program. They think outside the box and the sky is the limit. They use their gut to drive their answers to problems and give instructions in sound bites.

The psycho-geometrics tool allows you to quickly assess your staff's general personality traits so that you can give them tasks suited to their profiles. It allows you to understand the needs, desires, and expectations of employees. It also allows you to understand what information and feedback they need and assign tasks commensurate with their skills, needs, desires, and expectations. It is not an in-depth tool, but it is a down-and-dirty analysis that allows you to quickly and efficiently allocate work to your staff in the best possible manner.

Being a Great Boss

I'm not saying I am a great boss, but rather, I aspire to be a great boss. If you really want to know if I am a great boss, I would have you ask professionals who have worked for me. By and large I suspect you will get a truthful and objective response, except for those who were high maintenance and required a cattle prod to get them to perform.

Throughout my career as a boss I have discovered a few realities of great bosses and great employees. Some I have learned on my own, others I have learned from both good and bad bosses and employees.

Talent Squared

Great bosses and employees want the same things from a workplace. They want freedom from what I call the 3Ms: management's meddling, mediocrity, and morons.

To prevent meddling, I am a macro-manager who hires people who can do the job and, ironically, don't need to be managed. A little secret I have learned as a manager is the best way to not have to manage employees is to hire people that don't need to be managed. It's pretty simple, really. Secondly, as a manager, my job is to insulate employees from upper management. I protect them from upper management's penchant to intrude on their day-to-day routine of productivity.

For example, I recently went through a merger, and management was trying to find ways to charge employees for company benefits, like company vehicles used to drive to clients and cell phone use, among many others. When I was asked what I thought, I replied that they needed to be very careful. Why? Because, I told them employees are very ingenious. If you try and ride on their shoulders they will find a way to roll it up, double it, and stick it back in management's ear. If you start charging for company vehicle use, smart employees will turn them in and rent cars at double the rate, take a 1,000-mile trip, and charge the company for all of it. In the case of cell phones, employees would turn them in. Yes, they would save the cost of $39.99 per month, but employees would be inaccessible during the day on the road.

Lastly, as a manager my job is, when necessary, to insulate my employees from internal interference so that they can perform as expected. The way to handle this is to play your cards close to your chest and let your employees know this. If you are doing things differently but better, it is best to keep a low profile and not advertise, as it often draws more attention. Just keep your mouth shut and listen to others spill their guts. Meanwhile, you are operating in a "productivity zone" while others muddle along.

Employees are also looking for a change and a chance. A change refers to the fact that many companies mirror the *Dilbert* comic strip. Many want to have bosses they respect and admire, rather than criticize and abhor. They are also looking for a chance to free themselves from red tape and actually provide a benefit to the company. What a novel idea. When you free employees up, productivity goes through the roof. And guess what? Everyone is happy.

When you come to the decision to hire an employee, it is because you have enough work for yourself and one other person. Mathematically speaking, this means that you have 16 hours of work a day that needs to get done. Your goal is to hire someone to get that extra eight hours of work done so you only have to work eight hours a day. The dilemma is that when you hire someone to do that eight hours of work, you also have to spend time training and managing them, which adds to the eight hours of work you need to perform in a day. As stated earlier, your goal is to hire someone who doesn't need to be managed so that both of you can work eight hours a day.

Better yet, hire someone who is not only low maintenance, but maintenance free. That is, they add and multiply their time doing things way beyond their job description rather than subtract and divide from your time.

Good employees are looking for these types of bosses, who are often hidden in micro-cultures within companies. These employees often follow these bosses when they move up the food chain to sustain these types of productive work environments. You are creating a productive environment, often in the middle of a mediocre environment.

Comments you hear from these employees regarding great bosses include statements like:

- "He trusted me"

- "He insulated me from the bureaucracy" and
- "He never got in the way."

Traditional controls for these employees are irrelevant; trust requires no paperwork. Hence, the best way to manage is to hire employees who don't need to be managed. For the manager, the best way to manage is to not have to manage.

This sets you out as a great boss rather than an everyday ordinary boss.

Ordinary Boss	Great Boss
Offers good jobs at competitive wages and benefits.	Offer exceptional environment, including the opportunity to be exceptional. Knows that his/her employees are in a constant demand and so sets out to create a workplace that is "magnetic."
Has clear rules and ethics.	Has few rules but high standards. One-sentence summary, "Be big about the petty stuff and petty about the big stuff."
Understands how to function within the bureaucracy.	Knows how to function outside the bureaucracy.
Has answers.	Has questions. Understands that letting employees figure out the answers is more important than the answer itself.
Buys employees' time and effort	Buys help.
Allows you to keep growing.	Kicks you up to the next rung in your personal ladder of evolution.

When a work environment like this is created for employees, they:

- Never let themselves do second-rate work.
- Jump up to help when you are in a bind.
- Are their own worst critics.
- Dream your dream.
- Aren't like children.
- Watch your back.
- Have at least one skill superior to their boss's.
- Tap you on the shoulder and ask what you can hand off to them.
- Tell you what is needed, rather than you telling them.
- Have standards that lift us all up.
- Have an emotional attachment to their customers.
- Perform best in the midst of problems.
- Crystallize information.
- Show you the possibilities.
- Want to be tested.
- Are confident and seek measurement.
- Are entrepreneurial.

Can you imagine a boss who creates this kind of work environment or an employee working for a boss like this? Pinch me: it must be a dream. But it isn't, it is everyday at work. Wow, I can almost hear the superior productivity from here.

Great Bosses Don't Just Hire Employees, They Acquire Allies

As you travel through your career, you meet all kinds of people. You identify those who are leaders and those who are followers, those who are doers and those who need to be told what to do, those who are movers and shakers and those who need to be moved and shaken, and those who are smart and those who are, well, not so smart.

When it comes time to hire someone, you have identified those who would be an asset to the team rather than a liability. We've all made mistakes and learned afterward that we have hired the wrong person. As you progress through your career travels, it is a good idea to maintain a list of the ones who you would like to hire and keep their resumes and contact information, tracking them through their various employers. So, when the need arises, you can move quickly and fill the holes that need to be filled. These people are not really employees who work for you, but allies. They help you get in there, roll their sleeves up, and build the trains that need to be built to run the SH&E organization. As you move along in your career, they are the "traveling squad" that moves with you, each time taking a step up the food chain. This makes for a far more productive machine where personalities are known entities and the cohesiveness is at its apex.

Think about it. Which would you rather be? A great boss or a mediocre boss? A great employee or a mediocre employee? The difference is muddling through a series of jobs the rest of your life or experiencing great opportunities and solving problems for the rest of your career.

On Leadership

The Challenge of Leadership

Leadership is a hot topic among many SH&E professionals today. We Americans have a love affair with leadership. Go into any book store and you will find almost as many how-to books on leadership as you will find on 10-step programs to lose weight—and you all know how seriously we Americans take our waistlines!

The reason for our national obsession with leadership is, I suspect, two-fold.

First, as a free people, we believe that leadership is not a status granted to a few on the basis of birth, but a characteristic open to all on the basis of effort. Americans want to lead because we believe we can lead. Ironically, our longing to lead is an essentially democratic trait.

And, second, as a fortunate nation, we have often, since the days of Washington, Adams, Jefferson, and Madison, enjoyed leaders who have been the envy of other countries around the world. Americans want to lead because we know that leaders can change our communities, nation, and the world for the better. We prize leadership because we remain a fundamentally optimistic people.

I have found no better definition of leadership than the one recently put forward by James MacGregor Burns, a distinguished historian, who described leadership as a, "commitment to values and the perseverance to fight for those values." Leadership, at its most effective, is grounded in both principle and pragmatism.

Another historian, Garry Wills, drives home this insight in his book on leadership, *Certain Trumpets*. The most important quality of leaders is so obvious, Wills tells us, that we sometimes overlook it. Leaders have followers. "It is not the noblest call that gets answered," Wills reminds us, "But the answerable call."

Principles are essential, but stripped of a practical way to achieve them, they reduce all too quickly to impotence, just as pragmatism, unless harnessed to conviction, can slip all too easily into cynicism.

Leadership requires balancing an idea about the way the world should be with an understanding of the world as it really is represented. This is as true for nations as it is for individuals. The United States is the richest and most powerful nation in human history, but there have been rich and powerful nations before. What is different—and what explains, I believe, our ultimate victory in the forty-year struggle called the Cold War—is that the world trusts America and trusts us because of our values.

The Importance of Character

When my boss debates with me regarding why we need to implement these SH&E initiatives, I respond with a question: "Do you not litter because it is wrong, or because it is against the law?" If my boss responds, "Why, of course, because it is against the law," I say thank you and see you later. If he responds by saying that it is wrong, I quickly add that the same test should be applied to what we are about to do in this venture we are about to embark on.

Where did I get these values that guide the way I behave today? I got these values at home while I was growing up, instilled by my father and my mother. It mattered less what they said and more what they did. I imitated what they did more than what they said.

Unfortunately, many people today believe character is not as important as it was many years ago. I can recall being taught in elementary school that the main reason Abraham Lincoln was elected was because of his platform on character. Remember Honest Abe, who walked five miles trying to find the person who dropped a dollar on the road? Who would go to that trouble today? However, I submit to you that there is no more essential aspect of any person. A good practical definition of character is "what you do when no one else is looking."

Character is made up of those things that form us and give us meaning, direction, and depth. Character includes traits like honesty, integrity, courage, and munificence. Yet many see these as outdated and ill-suited for today. These same people think that all that is needed for success is talent, personality, and charisma, but history illustrates otherwise. Our founding fathers believed that these attributes formed the cornerstone of life leading to success and happiness. These character attributes led to freedom from evil, rather than freedom to commit evil.

When we as a society remove character as the core principle in the value system, heroes like athletes, musicians, and unethical business executives quickly replace the heroes of old. Remember when people like John Wayne and Dwight D. Eisenhower were hero figures? People with this type of character are rarely found in today's world. They have been replaced by bottom-line-oriented, self-centered, amoral folks who are clearly me-first. They espouse the thought, "I got mine, you get yours, at whatever the cost."

There is the common assumption that character is formed before adulthood. Character can, however, be built at any age by displacing self-interests with character attributes as core principles of a character-based value system.

Character Starts Within

In SH&E, intestinal fortitude comes from making the right decisions even though the immediate result may be unfavorable. When I first took an SH&E management position, an incident occurred on a weekend. Several counseled that the infamous OSHA-Recordable could be averted if only we interpreted the OSHA standard a certain way. It was decided to start off with character and integrity. Instead of taking the easy route, we took the high road. Our incident rate went up initially, but eventually, with focused program implementation, it declined. A promise was made that we were going to do this right, not by interpreting the standard in our favor, but to what the real intent of the standard was. It hurt the statistics initially, but paid off in big dividends to me, my department, and my management. It was important to be uncompromising, period. There was a standard and we stuck to it. We could all sleep at night knowing that an OSHA audit would reveal no covert actions on our department or my company.

As you walk down this narrow road, it starts off difficult, being challenged at every turn. Everyone seems to be trying to force you off the road onto side streets where you will eventually get lost. But after a while, people begin to understand your character and quit trying to get you off track. You become the standard by which everyone measures the truth. As time goes on, it becomes second nature rather than having to stop and think what is right and what is wrong. It is a wonderful way to live. You are free from worry and fear, knowing you are standing on the high road.

When something goes wrong and it is my fault, I admit it is my fault. How novel! Many people today, when things go wrong, are the first to blame someone else, exclaiming, "It didn't happen on my watch." Admitting your mistakes make the times of basking in glory even more satisfying. People know that you took hits along the way but you persevered and succeeded anyway.

These are all things that you do and say. There are also things that you do not do and say. You do not backbite your boss or other coworkers. It builds loyalty from those around you, knowing that as soon as they leave you will not participate in idle gossip about them. You defend those who are not present rather than join the crowd of backbiters.

Once you build and sustain character, everything falls into place. You know you are conducting your life as you should, treating others like you would want yourself to be treated, and being consistent throughout. Be careful: as you look around, you may find that others are looking to you as a leader.

Speak Like a Leader

Speaking like a leader will require you to think before you speak. By using very simple words, our vocabulary is imprecise, uninteresting, and unsophisticated.

Do you say:

- "Put up with" or "Tolerate"?
- "Risk" or "Jeopardize"?
- "Prove" or "Substantiate"?
- "Worried" or "Apprehensive"?

When speaking, which one of the words listed above would you choose, the simple one or the long one? If you chose the simple one, it may explain why your SH&E meetings are boring the employees. If you chose the long one, it may explain why employees appear more attentive.

As an SH&E professional, I do my best to make my presentations lively, entertaining, and articulate. I stop short of being esoteric. There are few occasions for such speaking–perhaps in the ivory tower of educational institutions and legal debates, but that is all.

Ask yourself four questions:

- Is it better to use short, simple words or longer, complex words?

- If you choose to use these words, how do you avoid pretentiousness?
- Do leaders really speak with a better vocabulary than most people? And if so, what terms do they use?
- Why is a good vocabulary important?

From listening to leaders and politicians speak, I found that they used more articulate speech. As a result, they are perceived as more intelligent communicators. After all, they are in positions that require great skill in influencing and guiding others.

To answer the questions posed above:

Is it better to use simple or more complex words?

If you keep it too simple, your vocabulary will eventually deteriorate to a child-like unsophisticated level and lose your effectiveness. For example, the word, "get" has fifty meanings and the word "thing" has more than twenty-five meanings in the thesaurus. Here are a few ways these words can be used:

Get:

- Did the SH&E manager *get* (purchase) the new fire truck?
- We *got* to (arrived at) the scene of the fire at 10:15 p.m.
- I *got* (received) the email yesterday.
- The industrial hygienist *got* (obtained) the monitoring data.
- Did the plant manager *get* (understand) what I meant?

Thing:

- I need to discuss two *things* (budget items) with you.
- Let's clarify this *thing* (issue) before it becomes a problem.
- We have three *things* (topics) to discuss during the SH&E meeting.

The illustrations above clearly show that by using only simple words, your vocabulary is imprecise, uninteresting, and dull. Using a more enriched vocabulary makes people want to listen to what you have to say and how you are going to say it. Of course, this assumes that you have the knowledge and experience to provide the substance of the answer. Your vocabulary is your tool to influence, just like a fault tree is a tool to find the root cause.

Using the words at the beginning of this section, let's put this to use and see which is more engaging.

- The SH&E manager was *apprehensive/*worried about the operability of the fire monitor.
- You could *jeopardize/*risk your SH&E by not wearing your SH&E glasses.
- The plant manager does not *tolerate/*put up with any opposition to his program.
- The reporter checked his facts carefully in order to *substantiate/*prove his statements were accurate.

It is clear to see which word adds more meaning to the statements above. These words, however, are not so high level that they are esoteric.

If you choose to use these words, how do you avoid pretentiousness?

Remember your high school or college English classes? Do you prefer to forget them, or did you sleep through them? Let's have a short history lesson. French entered the English language about 1066 when William the Conqueror invaded Britain. For 300 years, French was the official language of England, and we have both the French and Anglo-Saxon synonyms for many expressions.

In fact, today's English language has more synonyms than any language in the world. Given the enormous choice of words, how do we know which ones to use? By looking at the speech patterns of articulate leaders, we can draw conclusions regarding which words to use and create a simple system illustrating the least and most desirable words to use. Now, let's take a look at word origins:

Level 1	get rid of (ME)	stubborn (ME)	think (OE)	end (OE)
Level 2	eradicate (L)	obstinate (L)	contemplate (L)	terminate (G)
Level 3	extirpate (L)	refractory (L)	cogitate (L)	abrogate (L)

ME = Middle English; OE = Old English; L = Latin; G = Greek

Level 1 is the level most of us use in everyday speech, relying on simple words that are typically Anglo-Saxon. Level 2 words are used commonly by the media and leaders. Many of these words are derived from French, Latin, and Greek words. Level 3 words are unusual words, non-Anglo-Saxon-based, that would only be found in scholarly research journals or formal documents, and tend to be very pretentious. Leaders avoid them and focus on using the more intelligent words of Level 2 instead.

Do leaders really speak with a better vocabulary than most people? And if so, what terms do they use?

In my observation, I have found that leaders indeed use a more stimulating vocabulary. Leaders mix the articulate with simple expressions to balance their speech. They also use commonly understood words that are not commonly spoken but that are more stimulating, more authoritative, and more precise.

Why is a good vocabulary important?

The key reason for having good spoken vocabulary skills is *not to impress others*. Rather, it is to *influence others*, either in a public setting or a private meeting. By having a good vocabulary, we will be able to express our ideas more clearly. The clearer we are, the more credible we become. By becoming more credible, our ability to influence others increases. So, if you really want that extra $10,000 in your budget, speak more intelligently.

Obviously, when we speak in public at conferences and the like, our spoken vocabulary is critical. *What* we say is certainly the most important part of our speech, but *how* we say it is also critical. Consider standing in front of an audience in an old, untailored suit. Will you feel differently? Do you think the audience will perceive you differently? If something as simple as clothes will influence your perception as credible, so will your vocabulary.

The business community clearly recognizes and rewards the use of a higher level vocabulary. Just try to get ahead with a simple and bland vocabulary. Does your company president have a good speaking vocabulary? I'll bet the answer is yes! If you want to move up it is not a certainty, but it will most certainly help.

So, whether you are trying to convince your company president to give you $10,000 more for your annual budget or you are trying to convince Joe Operator to follow the Lock-Out/Tag-Out rules, an engaging speaking vocabulary is important in influencing others to heed your request. Whether you are presenting at a conference or chapter meeting, an engaging vocabulary will surely help you convince others to listen to you and not fall asleep.

The Significance of One

Every organization of which I have been a member has people who shift their accountability to others and don't vote. Remember the old saying, "There are three kinds of people: those who make things happen, those who watch things happen, and those who wonder, 'What happened?'"

In our overpopulated world, it is easy to underestimate the significance of one. With so many people, most of whom seem so much more capable, more gifted, more prosperous, and more important than me, who am I to think that my part amounts to much? However, that's the way most people think, and I dare say, SH&E professionals as well.

One school of thought is that we need to encourage people to vote and make a difference so that their voice can be heard resoundingly. If they surrender that power, then they deserve the second school of thought—don't complain about what is being done. In the second school of thought, I have found three characteristics are prevalent: apathy, ignorance, and indifference. When I ask people what they think on a particular topic they say, "I don't know and I don't care one way or the other."

What if Patrick Henry didn't say, "Give me liberty or give me death," but, "I don't know and I don't care, one way or the other."? What about Henry Ford, Martin Luther King, Walt Disney, Martin Luther, Winston Churchill, Jackie Robinson, Irving Berlin, Abraham Lincoln, George Patton, and Dwight D. Eisenhower?

If you don't believe in the significance of one consider the following:

- In 1645, one vote gave Oliver Cromwell control of England.
- In 1776, one vote gave America the English language as the national language instead of—holy mackerel—German. I think numbers were similar for Texas.
- In 1845, one vote brought the state of Texas into the union. Some are still pondering this move.
- In 1868, one vote saved President Andrew Johnson from impeachment.
- In 1875, one vote changed France from a monarchy to a republic.
- In 1923, one vote gave Adolf Hitler control of the Nazi Party. That should send chills up your spine!
- In 1941, one vote saved the Selective Service System just twelve weeks before Pearl Harbor was attacked. Yikes!

You may say to yourself, back then it was easy to stand out in a crowd. There were fewer people. There are still fewer people today who stand up for what is right and true, especially in SH&E. If we don't go along, we risk losing our jobs. My opinion is, I was looking for this job when I found it, and I can surely find another, especially if my employer chooses to do things unethically or illegally.

Never underestimate the power of one, especially if you are standing on the facts and the truth. These are hard attributes to overlook. Before you allow yourself to toss this aside and say, "All that's for somebody else, how much difference can I make, anyway?" Take a risk. Sacrifice yourself. It is the stuff leaders are made of. Leaders make a difference. Hillel once said, "If not me, who? If not now, when?"

I came across a quotation by Edward Everett Hale that really speaks to this. It is as follows:

I am only one, but still I am one. I cannot do everything, but still I can do something; and because I cannot do everything, I will not refuse to do something I can do.

I challenge you to ask yourself, "What should I be doing?" You alone can make a difference—a big difference. The question is, will you?

Success Tips

Compelling Communications to Upper Management

1. The Written Word
    - Be clear, be brief, and use short sentences.
    - Avoid the fog.
    - Write from their point of view.
    - Seize the reader's attention.
    - Have the data available.
2. Your Verbal Presentation
    - Perception differences. What do they understand?
    - Overload. Wait for the moment.

- Speed of listening. Some listen faster and want you to jump to the conclusion, others require you to slow down to capture the details.
- Timing. Wait for the workload presents the environment for your boss to say "yes."
- Trigger words. "Compliance," "risk," "exposure," etc. are example trigger words that upper management listens for.
- Semantics. Make your arguments relevant to the listener.
- Not listening for attitudes. Make sure your message doesn't fall on deaf ears.
- Listening. Don't just think of your reply or retort, listen so you can digest what is being asked.

3. PowerPoint Presentations: General Guidelines

- Don't blindly rely on the automatic graphic formatting provided by Excel or PowerPoint! Look at what you are displaying and choose the form that best illustrates your point.
- Strive to make large data sets coherent. This is particularly true when we are looking at accident causes that have multiple small percentages of certain accidents.
- Encourage the eye to compare different data. Make the data relevant by comparing to similar data sets over years or similar projects.
- Representations of numbers should be directly proportional to their numerical quantities. The numbers should accurately reflect the numerical value in a pie chart or a bar graph.
- Use clear, detailed, and thorough labeling. Don't make the reviewer interpolate or wonder what the label means.
- Display the variation of data, not a variation of design. The variation must match the trends and provide additional data that explains more detail that the other chart.

- Maximize the data to ink ratio – put most of the ink to work telling about the data! Invest in the picture not the number or table.
- When possible, use horizontal graphics: 50% wider than tall is usually best. This allows the reviewer to see the span the data includes, like months or years.

4. Edward Tufte's remarks on information presentation
    - Visual reasoning occurs more effectively when relevant information is shown adjacent in the space within our eye-span.
    - This is especially true for statistical data where the fundamental analytical act is to make comparisons.
    - The key point: "compared to what?"

Personnel Development and Delegation

- Trusting your staff
- Avoid seeking perfection
- Giving effective job instructions
- Recognizing the talent and ability of others to complete projects
- Recognizing skills
- Following up on progress
- Praising the efforts of your staff
- Avoiding reverse delegation
- Don't make delegation an all-or-nothing proposition
- Supporting your employees
- Delegating to the lowest possible level

Appendix A: Schools with SH&E Programs

ABET/NAIT Accredited Schools

Central Maine Community College, Auburn, ME
Construction Safety and Health (AS)

University of Central Missouri, Warrensburg, MO
Occupational Safety and Health (BS)

Indiana University of Pennsylvania, Indiana, PA
Safety Sciences (BS, MS)

Marshall University, Huntington, WV
Safety Technology Occupational Safety Option (BS)

Millersville University of Pennsylvania, Millersville, PA
Occupational Safety & Environmental Health (BS)

Murray State University, Murray, KY
Occupational Safety and Health (BS), Occupational Safety and Health (MS)

Oakland University, Rochester, MI
Occupational Safety and Health (BS)

Pennsylvania State University, University Park, PA
Industrial Health and Safety (BS)

Rochester Institute of Technology, Rochester, NY
Safety Technology (BS)

Trinidad State Junior College, Trinidad, CO
Occupational Safety and Health Technology (AAS)

West Virginia University, Morgantown, WV
Safety Management (MS)

Delgado Community College, New Orleans, LA
Safety & Health Technology (AS)

Indiana State University, Terre Haute, IN
Safety Management (BS)

Jackson State University, Jacksonville, AL
Industrial Technology (BS), Hazardous Material Management (BS), Occupational Safety/Health Technology (BS)

Nicholls State University, Thibodaux, LA
Safety Technology (AS)

North Carolina A & T State University, Greensboro, NC
Occupational Safety & Health (BS)

Owens Community College Toledo, OH
Environmental Technology (AAS)

Southwestern Oklahoma State University, Weatherford, OK
Industrial Technology (BS), Environmental Technology

University of Wisconsin-Platteville, Platteville, WI
Industrial Technology (BS), Occupational Safety Management

Other Universities Offering Safety Curricula

California State University, Northridge, CA
Environmental & Occupational Health (BS), Environmental & Occupational Health (MS)

University of California—Berkley, Berkeley, CA
MPH, Industrial Hygiene track

California State University, Fresno, CA
Environmental Sciences (BS)

UCLA School of Public Health, Los Angeles, CA
Environmental Health, Science, Industrial Hygiene (MPH, MS, Ph.D.)

Oregon State University, Corvallis, OR
Design and Human Environment (MS, MA), Design and Human Environment (Ph.D.)

University of Washington, Seattle, WA
Civil Engineering (BS), Environmental Engineering (MS)

Colorado State University, Fort Collins, CO
Environmental Health (BS), Environmental Health (MS, Ph.D.), Environmental Health–Ergonomics (MS, Ph.D.)

Boise State University, Boise, ID
Environmental Studies (BS), Environmental Health (BS), Environmental Health (MHS)

Montana Tech, Butte, MT
Industrial Hygiene (ABET/ASAC Accredited) (MS)

Southern Arkansas University, Magnolia, AR
Industrial Technology (AAS), Industrial Technology (BS)

East Central University, Oklahoma, Ada, OK
Environmental Health Science (BS)

University of Oklahoma Health Science Center, Oklahoma City, OK
Industrial Hygiene (MS, Ph.D.)

Texas Tech University, Lubbock, TX
Industrial Engineering (BS, MS, Ph.D.)

Lamar University, Beaumont, TX
Industrial Engineering (BS), Industrial Technology (BS)

University of Houston–Central Campus, Houston, TX
Industrial Engineering (MS, Ph.D.)

The University of Alabama College of Engineering, Tuscaloosa, AL
Industrial Engineering (BSIE, MSIE)

Embry-Riddle Aeronautical University, Daytona Beach, FL
Safety Science (BS)

University of South Florida, Tampa, FL
Occupational Safety (MPH)

University of Georgia, Athens, GA
Environmental Health (BSEH)

University of Louisiana at Lafayette, Lafayette, LA
Industrial Technology (AAS, BS)

Tulane University School of Public Health & Tropical Medicine, New Orleans, LA
Occupational Health; Environmental Management (MPH)

Mississippi Valley State University, Itta Bena, MS
Environmental Health (BS)

Illinois State University, Normal, IL
Environmental Health (BS)

Scott Community College, Bettendorf, IA
Health, Safety and Environmental Technology (AAS)

University of Minnesota, Minneapolis, MN
Environmental Health (MS, MPH, Ph.D.) (Specialty in Industrial Hygiene)

Missouri Southern State University, Joplin, MO
Environmental Health (BS)

University of Wisconsin–Eau Claire, Eau Claire, WI
Environmental & Public Health (BS, MS)

Salisbury University, Salisbury, MD
Environmental Health Science (BS)

Western Carolina University, Cullowhee NC
Environmental Health (BS)

Western Carolina University, Cullowhee, NC
Engineering Technology (BS)

Benedict College, Columbia, SC
Environmental Health Science (BS)

University of South Carolina, Columbia, SC
Environmental Industrial Hygiene, Environmental Quality, Hazardous Material Management (MS, Ph.D., MPH, MSPH)

Old Dominion University, Norfolk, VA
Environmental Health (BS)

Eastern Kentucky University, Richmond, KY
Public Health–Environmental Health Sciences (MS)

East Tennessee State University, Johnson City, TN
Environmental Health (BSEH, MSEH)

Bowling Green State University, Bowling Green, OH
Environmental Health (BS)

Ohio State University, Columbus, OH,
Industrial Engineering (Ergonomics) (MS, PHD), Industrial Engineering (Occupational Safety & Ergonomics) (MS)

Ohio University, Athens, OH
Environmental Health (BSEH)

Wright State University, Dayton, OH
Environmental Health Sciences (Ph.D.)

University of Massachusetts–Lowell, Lowell, MA
Work Environment Policy (BS, MS, Ph.D.)

Worcester Polytechnic Institute, Worcester, MA
Fire Protection Engineering (MS, Ph.D.)

Richard Stockton College of New Jersey, Pomona, NJ
Environmental Health (BA, BS)

Rochester Institute of Technology, Rochester, NY
Safety Technology (BS), Environmental, Health and Safety Management (MS)

Additional Reading

Attitude and Balanced Living

Carnegie, Dale. *How To Win Friends and Influence People*. New York: Simon & Schuster, 1936.

Covey, Stephen R. *First Things First: To Live, to Love, to Learn, to Leave a Legacy*. New York: Simon & Schuster, 1994.

Covey, Stephen R. *Principle-Centered Leadership*. New York: Summit Books, 1991.

Covey, Stephen R. *The 7 Habits of Highly Effective People*. New York: Simon & Schuster, 1989.

Durst, G. Michael. *Napkin Notes, on the Art of Living*. Chicago: Center for the Art of Living, 1988.

Dyer, Wayne. *Your Erroneous Zones*. New York: Funk & Wagnalls, 1976.

Frankl, Victor E. *Man's Search for Meaning: An Introduction to Logotherapy*. New York: Simon & Shuster, 1984.

Mackay, Harvey. *Beware the Naked Man Who Offers You His Shirt*. New York: W. Morrow, 1990.

Morley, Patrick M. *The Man in the Mirror: Solving the 24 Problems Men Face*. Brentwood, Tenn.: Wolgemuth & Hyatt, 1997.

Swenson, Richard A. *A Minute of Margin: Restoring Balance to Busy Lives*. Colorado Springs, Colo.: NAVPRESS, 2003.

Teresa, Mother and Joseì Luis Gonzaìlez-Balado. *Mother Teresa: In My Own Words*. Liguori, MO: Liguori Publications, 1996.

Waitley, Denis. *Seeds of Greatness: The Ten Best-Kept Secrets of Total Success*. Old Tappan, NJ: Revell, 1983.

Careers

Arden, Paul. It's *Not How Good You Are, It's How Good You Want to Be.* New York: Phaidon, 2003.

Capozzi, John M. *Why Climb the Ladder When You Can Take the Elevator?* New York: Villard Books, 1994.

Cialdini, Robert B. *Influence: How and Why People Agree to Things.* New York: Morrow, 1984.

Cottrell, David. *The Next Level: Leading Beyond the Status Quo.* Dallas, TX: Cornerstone Leadership Institute, 2006.

Kessler, Robin and Linda A. Strasburg. *Competency-Based Resumes: How to Bring Your Resume to the Top of the Pile.* Franklin Lakes, NJ: Career Press, 2005.

Kyne, Peter B. *The Go-Getter: A Story That Tells You How to Be One.* New York, Times Books, 2003.

Mackay, Harvey. *Swim with the Sharks Without Being Eaten Alive.* New York: Morrow, 1988.

Mandino, Og. *The Greatest Miracle in the World.* New York: F. Fell, 1975.

Maxwell, John C. *There's No Such Thing as Business Ethics.* New York, Warner Books, 2003.

Pollan, Stephen M. and Mark Levine. *The Total Negotiator.* New York: Avon Books, 1994.

Ries, Al and Jack Trout. *Positioning: The Battle for Your Mind.* New York: McGraw-Hill, 2001.

Ziglar, Zig. *Zig Ziglar's Secrets of Closing the Sale.* Old Tappan, NJ: F.H. Revell Co., 1984.

Management and Leadership

American Management Association. AMA *Survey on Corporate Downsizing, Job Elimination, and Job Creation: Summary of Key Findings.* New York: American Management Association, 1996.

Baldoni, John. *180 Ways to Walk the Leadership Talk*. Dallas, TX: Walk the Talk Co., 2000.

Bazerman, Max, H. *Negotiating Rationally*. New York: Maxwell Macmillan International, 1992.

Beck, Nuala. *Excelerate: Growing in the New Economy*. Toronto: HarperCollins, 1995.

Beck, Nuala. *Shifting Gears: Thriving in the New Economy*. New York: HarperCollins, 1995.

Bennis, Warren G. *On Becoming a Leader*. Cambridge, MA: Perseus Pub, 2003.

Bose, Partha. *Alexander the Great's Art of Strategy: The Timeless Leadership Lessons of History's Greatest Empire Builder*. New York: Gotham Books, 2003.

Burris, Daniel with Roger Gittines. *Technotrends: How to Use Technology to Go Beyond Your Competition*. New York: HarperBusiness, 1993.

Colan, Lee. *Sticking to It: The Art of Adherence*. Dallas, TX: CornerStone Leadership Institute, 2003.

Collins, Jim. *Good to Great: Why Some Companies Make the Leap and Others Don't*. New York: Collins, 2001.

Cottrell, David and Eric Harvey. *The Manager's Communication Handbook*. Dallas, TX: CornerStone Leadership Institute, 2003.

Cottrell, David and Mark Layton. *175 Ways to Get More Done in Less Time*. Dallas, TX: CornerStone Leadership Institute, 2004.

Cottrell, David and Mark Layton. *The Manager's Coaching Handbook*. Dallas, TX: CornerStone Leadership Institute, 2002.

Cottrell, David, Ken Carnes, and Mark Layton. *Management Insights*. Dallas, TX: CornerStone Leadership Institute, 2004.

Cottrell, David. *Listen Up, Leader!* Dallas, TX: CornerStone Leadership Institute, 2000.

Cottrell, David. *Monday Morning Leadership: 8 Mentoring Sessions You Can't Afford to Miss*. Dallas, TX: CornerStone Leadership Institute, 2002.

Dauton, Dale. *The Gifted Boss: How to Find, Create, and Keep Great Employees*. New York: Morrow, 1999.

Dellinger, Susan. *Psycho-Geometrics: How to Use Geometric Psychology to Influence People*. Englewood Cliffs, NJ: Prentice Hall, 1989.

Dodge, Brian and David Cottrell. *Becoming the Obvious Choice*. Dallas, TX: CornerStone Leadership Institute, 2001.

Fifer, Robert. *Double Your Profits in 6 Months or Less*. New York: HarperBusiness, 1993.

Garofalo, Gene. *Sales Manager's Desk Book*. Englewood Cliffs, NJ: Prentice Hall, 1996.

Goldratt, Eliyahu M. and Jeff Cox. *The Goal: A Process of Ongoing Improvement*. Great Barrington, MA: North River Press, 1992.

Harmon, Frederick. *The Executive Odyssey: Secrets for a Career Without Limits*. New York: Wiley, 1989.

Harvey, Eric and Al Lucia. *Walk the Talk and Get the Results You Want*. Dallas, TX: Walk the Talk Co., 1995.

Harvey, Eric and Paul Sims. *Nuts'nBolts Leadership*. Dallas, TX: Walk the Talk Co., 2003.

Harvey, Eric and Scott Airitam. *Ethics4Everyone*. Dallas, TX: Walk the Talk Co., 2002.

Harvey, Eric. 180 *Ways to Walk the Recognition Talk*. Dallas, TX: Walk the Talk Co., 2000.

Harvey, Erik, David Cottrell, Al Lucia, and Mike Hourigan. *The Leadership Secrets of Santa Claus*. Dallas, TX: CornerStone Leadership Institute, 2004.

Hennings, Joel. *The Future of Staff Groups: Daring to Distribute Power and Capacity*. San Francisco: Berrett-Koehler Publishers, 1997.

Hummel, Charles E. *Freedom from Tyranny of the Urgent*. Downers Grove, IL: InterVarsity Press, 1997.

Jeary, Tony and David Cottrell. *136 Effective Presentation Tips*. Dallas, TX: CornerStone Leadership Institute, 2002.

Kepcher, Carolyn. *Carolyn 101: Business Lessons from The Apprentice's Straight Shooter*. New York: Fireside, 2004.

Kouzes, James and Barry Z. Posner. *The Leadership Challenge*. San Francisco: Jossey-Bass, 2008.

Kyne, Peter B. *The Go-Getter: A Story That Tells You How to Be One*. New York, Times Books, 2003.

McLean, Emily A.N. *The Polished Image: A Concise Guide To Your Personal and Corporate Image*. Toronto: Nairne Concepts, Inc., 1990.

Peter, Thomas. *In Search of Excellence: Lessons From America's Best-Run Companies*. New York: HarperBusiness Essentials, 2004.

Peters, Thomas and Nancy Austin. *A Passion for Excellence: The Leadership Difference*. New York: Random House, 1985.

Phillips, Donald T. The *Founding Fathers on Leadership: Classic Teamwork in Changing Times*. New York: Warner Books, 1997.

Porter, Henry. *Secrets of the Master Sales Managers*. New York: AMACOM, 1993.

Sokolosky , Valerie. *Monday Morning Leadership for Women*. Dallas, TX: CornerStone Leadership Institute, 2003.

Taylor, Harold. *Delegate: The Key to Successful Management*. New York: Warner Books, 1991.

Tufte, Edward R. *Envisioning Information*. Cheshire, Conn.: Graphics Press, 1990.

Tufte, Edward R. *Visual Explanations*. Cheshire, Conn.: Graphics Press, 1997.

Tufte, Edward R. *The Visual Display of Quantitative Information*. Cheshire, Conn.: Graphics Press, 2001.

Tukey, John W. *Exploratory Data Analysis*. Reading, MA: Addison-Wesley Pub. Co. 1977.

Tukey, John W. and P.A. Tukey. "Summarization: Smoothing; Supplemented Views" in *Interpreting Multivariate Data*, edited by Vic Barnett. New York: Wiley, 1981.

The Black Book of Executive Politics. New York: National Institute of Business Management, 1994.

Personal Growth & Personal Finance

German-Grape, Joan. *Ninety Days to Financial Fitness*. New York: Maxwell MacMillan International, 1993.

Johnson, Spencer. *Who Moved My Cheese? An Amazing Way to Deal with Change in Your Work and in Your Life*. New York: Putnam, 1998.

Maltz, Maxwell. *The Magic Power of Self-image Psychology: The New Way to a Bright, Full Life*. Englewood Cliffs, NJ: Prentice-Hall, 1964.

Schwartz, David J. *The Magic of Thinking Big*. New York: Simon & Schuster, 1987.

Tobias, Andrew P. *The Only Investment Guide You'll Ever Need*. San Diego, CA: Harcourt, 2002.

Vest, Herb D. and Lynn R. Niedermeier. *Wealth, How to Get It, How to Keep It*. New York: American Management Association, 1994.

Weihenmayer, Eric. *Touch the Top of the World*. New York: Dutton, 2002.